The Age of Power

By the same author

WASTE AND WATER

The Age of Power

J. H. STEPHENS

PETER DAVIES : LONDON

1974

Peter Davies Ltd
15 Queen Street, Mayfair, London W1X 8BE
LONDON MELBOURNE TORONTO
JOHANNESBURG AUCKLAND

First published 1974

SBN: 432 15725 3

Printed in Great Britain by
Richard Clay (The Chaucer Press) Ltd
Bungay, Suffolk

Contents

Illustrations

Between pages 88 and 89

Diagrams and Graphs

Tables

1 The bison and the lamprey

On my twentieth birthday, Hiroshima was devastated by an atomic bomb. The bomb had an explosive power equivalent to about twenty thousand tons of chemical high explosive; that is, around a thousand times greater than any comparable weapon known up to that time. A second bomb was exploded over Nagasaki, and the might and destruction of these two bombs ended the war with Japan. About seven years later (1952), the first hydrogen bomb was exploded. It increased the destructive power of explosions again by an order of a thousand, and bombs equivalent to fifty or a hundred million tons of chemical high explosive are now possible. Radioactivity released by such explosions is highly poisonous to living things. Destruction on a global scale of all human life is thus rather simple today, and the novelty of this possibility dominated political and public opinion for twenty years or so after the war, and was thrust into the public eye at the time of the Cuban crisis. Since that time it has become more popular to stress other global fears. There is some reason in believing that the rate of increase in population is the world's greatest threat, from which other anxieties derive: that natural resources will be exhausted, or that the world will become poisoned by the polluting wastes of modern industry and life.

Usage of energy is at the centre of this hubbub. Control of the atomic nucleus and the great store of energy therein was the key to massive destruction; but this feat has proved up to now to have been almost a side effect of the mainstream of competence in applying sources of energy to useful purposes. Energy is the food of modern life. The great variety and richness of mankind's activities – man's leisure, his transport, his industry – all depend on consuming energy in a multitude of devices which function as inanimate servants, as teams of horses, as tools for quite novel activities or to multiply many times over the efficacy of traditional ones.

Usage of energy is the common necessity where heat, or cold, are needed, or where useful mechanical work has to be done. Energy is consumed, but in small amounts, for what I call brain-like uses, for computers or communications for example, but in this book, whilst brain will not be formally denied, the subject is brawn, and brain-like uses if important qualitatively, fit in as a tiny proportion of the quantities (one morning's concentration equals half a peanut) purveyed by the industries which supply fuel and energy in their various forms.

Brawn, however, has its subtleties. There are fundamental physical laws governing the diversity of energy usage, and thrown into relief by it. They are overlaid by a richness of practical inventiveness, by the spur of competition in price and by demands for energy which, universally, have grown exponentially and continue to do so. Whatever changes the future may hold, they will be governed by the fact that energy will continue to be consumed if the character of civilized life is not to be destroyed.

Our subject is thus at the heart of things. To understand it, one needs a bird's eye view over a great landscape, peopled by many and diverse specialists. This book is an attempt to give such a bird's eye view. Now if one's bird is a hawk, he will report a landscape with rodents waiting to be caught; if he is a bullfinch, he will report a landscape with orchards. The diversities amongst energy specialists are perhaps greater, and there will be gaps, but the broad avenues of progress, trimmed of the abstruse, should be represented.

Thus a setting is created from which to understand the present and the future. Energy being the food of life, one might ask whether mankind will have emptied the platter in a few generations' time? Will the world's resources of fuel be dissipated almost as speedily as the bison disappeared from the North American plains? One year, herds which took all day to pass by; a few decades, virtual extinction. In the meantime, will the diet be too rich? King John reputedly died from a surfeit of lampreys. Mankind's thermodynamic digestion perhaps could prove equally sensitive to his diet of burned fuels, noxious exhausts and radioactive wastes. Eminent critics have suggested both of these possibilities, to which we return in later chapters, noting meanwhile that bison apparently are on the increase, and that medieval kings suffered many sorts of ghastly demise.

More optimistic possibilities also exist. The greatest is that energy of atomic fusion – the source of energy of the sun and of the hydrogen bomb – will eventually be commanded on a useful scale, so providing a benign and inexhaustible service, and so rendering obsolete most of the magnificent and gigantic artefacts which are featured in these pages. Again, the possibility is discussed later, in its place, as our theme develops from an inevitably humble start with a little indoctrination of scientific theory.

A PILL FOR THE READER

A simple explanation of the most important of the physical principles which underlie the utilization of energy is needed right at the beginning of the book, to provide the basis for understanding the chapters which follow. In other words, I have to put the pill on the outside and the sugar coating inside. The pill will be kept as simple as possible, consistent with ensuring the reader's easy digestion of the sugary centre, and I will not labour it with the pedantry of a textbook.

Work

The basic notion from which we start is that of work – mechanical work, measured by a force acting through a distance. One foot-pound of work has to be done to lift a weight of one pound vertically upwards through a distance of one foot.

Energy

When work is done, energy is expended. Thus energy can be measured in the same units as work, and the two are to some extent synonymous. But the concept of energy is wider than that of work. Energy is sometimes defined as 'capacity for doing work', which implies that it can be stored in some way, or can be in the process of being utilized or dissipated. Energy exists in many forms, one of which is mechanical work. This simple explanation and attempt at definition clearly derive from an engineer's line of thought, which is the type of explanation that we need here. Definition alone however is unlikely to be adequate, and understanding must rely as well on a certain knowledge of the physical nature of the universe, of which the two basic qualities are mass and energy.

Forms of energy

The simplest forms of energy are closely akin to mechanical work. Energy may consist of some physical system which has pressure, velocity or 'potential'. In a gravitational environment, height is equivalent to energy, and is the simplest example of potential energy.

Energy may also exist as heat; as chemical energy; as electrical energy; as light or electromagnetic radiation; as a magnetic field; or as energy of the atomic nucleus. Electrical energy is dynamic in its character; it cannot be stored (with perhaps minor exceptions), as can some of the other types mentioned, but has to be generated and utilized in one process.

There is a fundamental difference between atomic or nuclear energy and all the other types, including chemical energy. Nuclear energy is released by transmutation of mass into energy, when the structure of the atomic nucleus undergoes some change. Atoms may be considered as indestructible and unchangeable when we consider chemical energy; there is an energy of bond between the atoms of different elements which combine to form a compound: this is chemical energy; it can be released, generally as heat, when a bond is broken, or when a bond is made.

One other class of energy is that exerted by living things. The muscles of living animals, and indeed of man himself, gave rise to the notion of work, and 'mechanical work', defined above as a starting point, is a later stage of sophistication.

For practical purposes, these forms of energy reduce principally to three, namely electrical, heat or thermal, and mechanical energy.

Power

The rate at which energy is expended is called power. In the simple mechanical case, power is 'the rate of doing work'. To lift one pound through one foot in one second requires the same power as to lift one pound through ten feet in ten seconds, although ten times the quantity of energy will be expended.

Conversion of energy

Energy is indestructible. It can be converted from one form into another, and it can be dissipated, but it cannot be destroyed. This principle is known as the first law of thermodynamics, which states that in a closed system the total quantity of energy is constant. The law also states that when one form of energy is converted into another there is a fixed ratio between the quantities of the two forms.

But when heat is being converted, some of it is always wasted. This is a very important qualification, and the reader really needs to know why it is so. The explanation is in the Appendix. The Appendix also explains how energy is measured. I recommend both explanations; the figures quoted throughout the book will lose meaning without them. But I have been persuaded that a pill right at the beginning must be a tiny one, or the dose will be refused and with it the entire course of treatment. Ideally, I should conjure up a dancing girl or a comedian at this point, to explain the second law of thermodynamics. As second best, I leave the reader a choice, and hope he will have the temerity to tackle the Appendix next.

2 Capacity for doing work

As a starting point, how much energy is there, how much capacity for doing work, in the muscles of a man? According to one authority the work done by one man at the peak of physical fitness is equal to the output of one light bulb of 40 W burning continuously. One day's work is thus 960 Wh, which in an eight-hour day is equivalent to a rate of one-sixth of a horse-power (see Appendix and Table 9 for definitions of kilowatt-hour and horse-power). Another says, 'In his most vigorous age man can at best produce approximately the equivalent of 1 kWh of electrical energy over a period of 12 hours of strenuous effort and work.' A more convincing measurement still is found in the work of two researchers at the University of Washington, Seattle,* who measured the outputs of different individuals doing different types of work (on a sort of treadmill, cranking a wheel with two hands or working a lever to and fro) at different rates. The maximum rate of overwork was governed by the rate at which the heart beat. At 150 beats per minute, about the maximum, a man 25 years old and weighing 160 lbs, but in a sedentary job and not physically trained, consumed oxygen (measured by the heart rate in these experiments) equivalent to 22,900 ft lb per minute, and provided useful power to the treadmill of 5,500 ft lb per minute. He could do this for 15 minutes.

In the Imperial units in which these measurements were made, 33,000 ft lb per minute is equivalent to one horse-power, so for 15 minutes this medium-sized, sedentary young man was working at a rate of one-sixth of a horse-power (about 125 W) and his useful output over 15 minutes was about 30 Wh. His efficiency was measured; it was 24 per cent for this work measured simply as output divided by fuel, but taking a net figure (that is, making allowance for the energy he consumed when just standing quietly) it was 32 per cent, and the efficiency of conversion of the extra energy used to meet the peak of the work was 62·5 per cent. Other modes of working were much less efficient (cranking, 10½ per cent gross, working a lever, 8½ per cent gross). So the gross efficiency of the human animal is about equal to a gas turbine or a diesel engine, but not as high as a modern steam-turbine-alternator set, which can reach 35 per cent or so.

* B. W. Owens and G. Latta, 'On the measurement of human power', *The trend in engineering*, Quarterly Journal of the University of Washington, College of Engineering, January 1966.

Adding a little guesswork to the experiment, if the peak output could be repeated after 15 minutes' rest over an eight-hour day, then the day's useful work would be 480 Wh, say half a kilowatt-hour, which is less than the other two estimates. Making every allowance one can on these figures, a man's own physical efforts remain pitifully puny.

Throughout historical times, he has augmented them by the work of domestic animals – the horse, the ox or the elephant. One of James Watt's accomplishments was to define the value of one horse-power. His purpose was essentially similar to my own in this chapter. He needed to give a measure to the power of his engines that could be easily understood in every-day terms, and which could be applied to his 'rotative' engine which he introduced in 1781. Engines which pumped water were compared by their 'duty' – the quantity of water the engine would lift through one foot for the consumption of a specified quantity of coal. This measure (which is not dissimilar from the heat rate used to compare performance today) could not be applied to an engine which turned a shaft, and so operated machinery. Watt did not sell his engines for a fixed price, but charged a premium calculated from the savings in coal compared with a Newcomen engine, so a new standard of comparison had become essential. He ascertained that a horse turning a mill walked 2½ turns a minute round a path 24 feet in diameter. He also assumed that the horse exerted an average pull of 180 lb, but it is not known how that figure was fixed. The 'rounded' answer of 33,000 ft lb per minute has been used unaltered ever since as one horse-power.

I tried to unearth results of similar character and quality to those of Owens and Latta, but applied to horses, and I hoped elephants too. The search proved fruitless in modern literature, but I found some interesting data in Edward Cresy's *Encyclopaedia of Civil Engineering*, published in 1861. Cresy records the measurements of noted French savants, such as Coulomb and Daniel Bernouilli, in measuring the work of both men and horses, at considerable length in a chapter called 'On mechanical agents or the first movers of machinery'. Very briefly, here are some results. The daily labour of a man was 1,728,000 ft lb – that is, equal to about one-ninth of a horse-power for 8 hours, or about two-thirds of a kilowatt-hour. In some types of work measured, a man's output dropped to half or less than half of that figure. The daily labour of a rower was about 10 per cent greater, say three-quarters of a kilowatt-hour. Not bad agreement!

Cresy states as a generality that a horse will travel 200 feet a minute while pulling with a force of 140 lb (about the same distance but a lower force than Watt's measurement). This makes his estimate of a horse-power 28,000 ft lb per minute as against Watt's 33,000. The 'tractive power' of a good wagon horse – by which he means its maximum pull – he gives as 140 kg. A wagon could travel 38–40 kilometres in a day of 8 or 9 hours on good level roads, he notes, but this speed is slightly more than 200 feet a minute, so presumably the wagon horse only pulls on average at substantially less than half of his 'tractive power'. Then follow extensive measurements of useful work obtained, over a period, for horses working various types of machinery; both the useful output of the machinery and the effort of the horses were measured, since the machinery had a mechanical efficiency of half in most cases, a quarter in one example. The object of the tests was to measure the 'useful effort' which takes in the inefficiency of the machinery as well as the work done by the prime mover. Converting the figures back to just the work done by the horses, the answers vary from 4 horse-power hours in a day's work of 4–6 hours, 6 horse-power hours for a day's work of 8 and one of 12 hours (with different machinery) and 10 horse-power hours in 6 hours, without a correction for the machinery. This last figure Cresy discards; he thinks it is too high. The output of a good wagon horse on the basis of these tests is thus at a rate of ½ to 1 of Watt's units throughout a day's work, so if we take a normal day's work which is a reasonable unit since the rate of working tends to be higher for a short day, and vice versa, then we can say that a man will supply about 0·5–1 kWh and a horse about 3–6 kWh, of physical work.

Until the industrial revolution, the only sources of energy with capacity for doing work were muscles, flowing water and wind. It can be argued that even the last two sources were not used very much until the first stirrings of that revolution were in the air, but there is no need to pursue such a fine point when the purpose is merely to compare, in terms of energy, the resources of the era of muscles with those of today.

As an example, take a farm of say, 800 acres (about 3 sq. km.) worked by manual labour and horses. It might have been fortunately placed and received some supplement of energy for special purposes, directly or as services. For example, say 150 years ago, corn could be ground in a windmill or a water mill; produce could, perhaps, be sent to market by water using the wind's energy for transport. By the mid-nineteenth century, changes would have occurred, with railways for transport perhaps. Steam ploughing was introduced on some favoured farms, but its main purpose was to make better progress rather than to replace men. The twin engines which hauled the plough needed a carter to feed them with water and coal.

Before the First World War, a tractor with a corn-cutting machine could cut a peak of 20 acres of corn in a day, but it was a rare sight at that time. The use of power from the internal combustion engines of tractors mainly, and also from electricity, progressed with changes in the actual methods of farming on the one hand, and with the development of tools and implements to use the power on the other. This progress illustrates one of the main themes that will recur later, by demonstrating that the availability of a source of energy does not itself promote development, but lack of it when it is wanted can be a serious impediment.

To return to the 800-acre farm. Depending on muscles, it would have about twenty men working long hours, and about twenty horses. In terms of daily useful energy that would be a maximum of about 20 kWh from the men and 120 kWh from the horses, say 25 to 35 MWh* per year. An interesting comment on the manpower figures is that men worked harder in those days, and generally they were able to do more physically, and for longer hours. The canals of Britain and most of its railways were built by manual toil. Much of Holland was reclaimed from the sea by manual toil. Men may have been puny, but they were persistent.

The same farm today would be yielding much more produce, but it would be more specialized. If it yielded say, half corn, half beef, four men would work it with, say four tractors of about 70 horse-power each, equipped with many implements and devices – to blow air for drying grain or hay, to drill, to roll, to harrow, to broadcast, to bale, to plough, to dig holes and tension wire for fences. There would also be a powered elevating or loading device (say 20 h.p.) of some kind, because bulk handling, in place of sacks, has proved to be an important labour-saving technique. And of course a combine harvester – of 80 or 100 h.p. – which could cut and thrash 20 acres of corn or more per day. There would maybe be a four-wheel-drive vehicle for transport around the farm, but neglecting transport, the farm would have around 500 horse-power available. The output of 'useful' energy of the men would be much lower than from the old-time farm, because they would mainly be operating machines. Machines are not adaptable; men are, and their intelligence becomes more important than their muscles. The corresponding output of energy of the machines on the farm might be about 200 MWh in a year. The daily figure is not really a good comparison, and even the yearly one can vary depending on the hours during which the machines will run. Output of energy will depend on what the job requires, and the energy is there if it is needed.

If the farm had a dairy herd of 70 or 80 cows, the dairy side would be electrically operated, with a milking parlour run by one more man (as against five milking by hand). Its consumption of energy would be small

* k means a thousand; M means a million; see Table 9.

in relation to its usefulness. Electricity might be used also for drying and loading grain. The cost of the electrical energy would be a small item, out of proportion to the job it carried out. The decision to install electricity, and with it expensive equipment, is the vital stage, and it is at that point that capital has to be spent. The same is true for the engine-powered devices, which are expensive to buy, and in the case of the combine will be used for only a few weeks a year, so its actual output of energy is very small compared with the job it does. Totting up the horse-power per acre or per worker gives an indication of greater competence, but can be misleading unless illuminated by knowledge and experience of what is going on. The same applies to other statistical arguments about energy.

Usage of energy in the home has changed in a similar manner to the way just described on the farm. Even heating, which strictly should be excluded from this discussion, has been influenced, because of more extravagant ideas about comfort and the use of hot water. Cooking and heating, the low-grade users of energy thermodynamically, will account for the biggest part of the energy consumed, but the horse-power for useful work will not be negligible, as can be checked in any home. Modern life may lack some of the spaciousness of former times, but today's gadgets are equivalent to a houseful of servants, or a stable-ful of beasts of burden. The greatest of these, and the one with the most far-reaching special effects, is the family motor car, which may commonly have around 50 or 100 horse-power under its bonnet.

The amounts of energy needed by industry today are so enormous and the industrial activities themselves are sometimes so new, that a simple comparison between old and new cannot easily be made. Horse-power per worker in industry can be as high as 300. In some industries – for instance, refining aluminium ore – a great deal of energy is obligatory; in others a high figure indicates advanced mechanization or automation.

Energy consumption per head nationally is another favourite statistic. If one compares consumption per head in olden times with today, the figures will probably include heat energy, which is better omitted from this discussion, because the unique advantage today is in the use of high-grade mechanical energy, and a lot of heat is wasted to get it.

Another point is that it is social motive and purpose that create the demand for energy. Although national standards of material wealth correlate statistically with *per capita* energy consumption, the fact of having the energy does not create the standard; energy consumption is a passive follower and not the *raison d'être* of the statistics.

Since today's standards and demands are our main interest at this stage of the book it is worth getting an idea of the scale of usage of energy – domestically, industrially and nationally. This is done in Table 11 (see Appendix); the figures quoted have been chosen to illustrate the gist of the text of this chapter (main sources of energy are in Table 10) and compare typical figures for human usage with the scale of natural thermodynamic phenomena. Probably most of the figures are on the rough side, but they give a comparative standard and a perspective which gives meaning to the energy statistics which occur from time to time throughout the book.

The tables give the bare bones that one needs. Otherwise there is no end to the statistics. They should be seen against a trend of rising world consumption. The figure of 80 per cent losses on a worldwide basis may cause some surprise, but if approximate, is authoritative and unchallenged. By reducing losses from 80 to 60 per cent therefore, one doubling of world consumption would be possible without increasing the rate of consuming fuels. Note also that ten lifetimes of physical work is the equal in energy of one year's consumption by an American, but bearing in mind it is heat and high-grade energy, mixed up together.

The thermodynamics of natural phenomena throw interesting light on these statistics. The sun is a furnace, fuelled by nuclear fusion, and exports energy by radiation. Recent research information is shown in simple form in Fig. 1 below. The earth is in net balance, and re-exports the energy it collects from the sun as infra-red radiation. About a third of the sun's incoming radiation is directly reflected off the earth. For the rest, very broadly the tropical zones gain heat and the arctic zones lose heat. The atmosphere and the ocean are together the working agents of a terrestrial heat engine, the cycle of which is all the phenomena of weather. This terrestrial heat engine has a low thermal efficiency in its day-to-day operations, but has interesting features such as great capacity to store heat in the oceans or in the polar ice caps. To melt the ice in Antarctica would need all the energy received from the sun for 2½ years; the same quantity of energy would raise the temperature of the oceans of the world by little more than one degree Centigrade*.

How the terrestrial heat engine works in detail is not well understood yet, which is one reason why some of the predictions made about the effects of future energy usage are somewhat uncertain. For example the conclusion one draws from the figures just quoted is that the heat reservoir which the oceans form on the earth's surface is so large that if the earth's heat balance were disturbed (if there were a net gain or loss of solar radiation) the effect would be masked for a long period and there would be little change in air temperatures on the earth's surface. This

* Seymour Tilson, 'The ocean', *International Science and Technology*, February 1966.

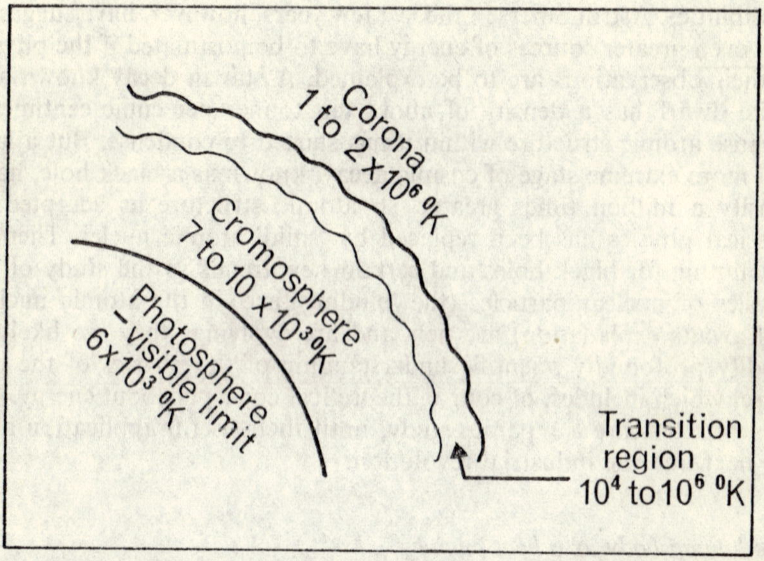

Fig. 1 The sun's energy is radiated mostly from the photosphere. Recent research has investigated the corona. The transition region is 10^3–10^4 kilometres from the photosphere. Energy flux in the corona in ergs per square centimetre per second is:

radiation	5×10^4
conduction	3×10^5
wind	4×10^4

(Source: Culham Laboratory.)

conclusion contradicts the predictions made in Chapter 6 where the increase in CO_2 in the atmosphere is discussed.

The scale of terrestrial thermodynamics is, in spite of man's newly acquired powers, so vast as to be difficult to comprehend. The tropical hurricane is probably the most violent amongst the diversity of weather's events. It forms over a tropical ocean, from which it extracts and vaporizes warm water at a rate of the order of a million tonnes a second. When the vapour condenses into liquid its latent heat is released and this is the source of energy of a hurricane, the magnitude of which is indicated in the Table 11.

Table 11 also shows that the transformation of mass into energy, which is the characteristic process of nuclear energy (Chapter 5), could, if carried to the ultimate of 100 per cent transformation, yield fantastic quantities of energy. This possibility was always considered to be the ultimate of which the laws of physics could conceive. The universe, and the physicist's scheme of explanation of its properties, offered no other possibilities. Astronomers in the last few years, however, have suggested that even greater sources of energy have to be postulated if the physics of their observations are to be explained. A star in decay known as a white dwarf has a density of about ten tonnes per cubic centimetre, because atomic structure within it has started to condense. But a later and more extreme stage of cosmic decay, known as a black hole, has a density a million times greater, so atomic structure as accepted by classical physics has been replaced by 'solid' atomic nuclei. Theories accounting for black holes and certain new trends in the study of the physics of nuclear particles (the 'binding face' in the atomic nucleus still awaits explanation) are new and are evolving; they are likely to modify profoundly scientific understanding of the physics of the universe, which includes, of course, theoretical concepts about energy. But these matters are a separate study, until their useful application after the next but one industrial revolution.

The human body as a heat engine

Since this chapter started on a human scale by discussing a man's output of energy, it can be concluded by relating his metabolism also to our theme. The ramifications of thermodynamic theory are endless; take the human body, and consider it as a heat engine. First, we must make a rather serious qualification, since the human body operates at constant temperature. Transfer of energy is achieved chemically, at one temperature, in a manner closely analogous to a heat engine, but at first glance apparently in contradiction of the second law.

Consider an internal combustion engine. Its cylinders take in a combustible mixture of fuel, burn the fuel at the high temperature and reject the burned gases through its exhaust to the sink. The cylinders are cooled by a fluid, generally water, which itself is cooled by heat interchange with the sink at the low temperature, and most of the heat rejected by the engine is lost through the cooling fluid.

Now the body. A digestible fuel (food) is taken into the stomach and burned (digested), thus adding energy to the system at a high chemical potential, the potential of biochemical bond being substituted at constant temperature for the potential of high temperature in the engine. The residual energy of low potential in the fuel is rejected by excretion. Circulation of the blood is in thermodynamic terms analogous to the circulation of cooling water in the engine, but again, since the temperature of the body is constant, energy is rejected through the lungs by virtue of the different energy potentials of chemical combination of blood arriving at the lungs (i.e. rich in carbon dioxide) and blood leaving the lungs (rich in oxygen). The carbon dioxide exhaled corresponds to the heat rejected by the cooling water of the engine.

The thermal efficiency of the body (perhaps one should call it thermo-chemical or thermo-biochemical efficiency) is about the same as that of a good heat engine, and subject, it would seem, to closely analogous limitations, though I have never seen them stated in a second law of thermo-biochemical theory. To describe the human body as a 'constant temperature heat engine' would appear to be accurate rather than bizarre. So where has a constant-temperature engine been devised in verification of this principle? Chemical methods of generating electricity are analogous – that is electricity from a battery – and by that method mechanical work can be done by driving an electric motor from the battery, all at one temperature. But there appears to be no direct link between chemical and mechanical energy, without either electricity, heat or a living organism as the intermediary. Another example of a similar kind is the fuel cell.

Returning to the work of Owens and Latta, whom I quoted at the beginning of this chapter, they put a different emphasis on the analogy. Body fuels are burned by consuming oxygen, they point out, thus:

> As the power output of the combustion engine is regulated by carburation, so is a man's working potential dependent on his capacity to transport oxygen and to burn the food he eats. If we continue the analogy, the protein elements of foodstuffs correspond to the structural material of the engine; water, to the lubricants; carbohydrates and fats are the fuel; and the vitamins correspond to the ignition sparks.

3 Thermodynamic sunrise

Muscles, wind and water lost their monopoly as the world's only sources of mechanical energy strictly to Savery, in practice to Thomas Newcomen. Other heat engines had been demonstrated – Hero's is the best known – but they were scientific curiosities. Savery's 'steam pump' (1698) was actually put to work, although it had very limited possibilities. Newcomen's engine was the significant prototype. The first one was installed at Dudley near Birmingham in 1712. The only use initially for the Newcomen engine was to pump water from mines, notably the Cornish tin mines. It was an expensive device, and its cost was justified only in special applications, but with experience it began to be improved.

In 1709 Abraham Darby introduced a successful method of obtaining iron by smelting iron ore with coke, making possible abundant supplies of iron. In 1717 Thomas Lombe set up the first true factory at Derby, where he employed 300 people in silk-throwing. His mill was water-powered, and had severe limitations. It was not until the steam engine made power available at any time that Lombe's prototype could develop. These innovations – the factory, the material and the source of energy – are considered to be the birth of the industrial revolution. Here were the basic requirements for industry in a modern sense: a steam engine and iron to make it with, and to make other artefacts with which to manufacture, mine and so forth.

There is another view also, of the birth of modern industrial society:

> The real failure of the ancients resided in their relative inability to make use of any source of energy or of any prime mover which was not composed of flesh and blood. Even the wind applied to the sailing ship was relatively useless to them, owing to the absence of the compass and the chronometer and the limited capacity of their ships for sailing into the wind. If I were to fix a date for the real beginning of the industrial revolution, I think I should put it at a time when the ocean-going sailing ship and the compass and the chronometer made technically possible the circumnavigation of the globe and its systematic exploration.*

James Watt's best-known invention, the use of a separate condenser, was made in 1765. In the Newcomen engine the cylinder itself was

* Lord Hailsham, from his opening address at the centenary celebrations of the Institution of Electrical Engineers, London, May 1971.

heated and cooled during each cycle; Watt devised a separate 'sink' – the concept discussed in the Appendix – so reducing enormously the waste of heat. The economy which this measure made possible increased the range of applications for which steam engines could be used. This in its turn encouraged the introduction of further improvements and the manufacture of larger, more ambitious engines, as well as more compact ones. The steam engine had evolved to the point where it had become a practical source of power. The factory too, as the basis of modern ideas on production, took a step forward with the completion of the Soho Foundry in Birmingham by Boulton, Watt and Sons in 1796. The foundry was years ahead of its time as an example of industrial production and organization.

The history of early steam cars and locomotives is not our subject here; suffice to record that the railway age was one result of these developments. The opening of the line from Stockton to Darlington in 1825 can be taken as its starting point. The fever of railway construction reached a peak in the mid-nineteenth century. By that time, however, two other discoveries of equal significance to Newcomen's had been made, but that significance did not become apparent until half a century or so later. This was the age of the reciprocating steam engine, mainly fuelled by coal but sometimes by wood, at its peak on the railways, but also used as the source of power throughout the rapidly developing factories, mines and other installations of the emerging industrial society. At sea coal-fired steam ships competed with sail, and on land the stage coach was superseded, but the horse remained pre-eminent for local transport.

In 1831 Michael Faraday discovered the principle of electromagnetic induction. He inserted a magnet into a coil of wire and generated, as he described it, 'a wave of electricity'. This phenomenon he explored in ten days of experiments 'so thoroughly that no one has since been able to add anything to his work in the discovery of fundamental facts'. The greatest of these discoveries was 'that an electric current could be generated in a wire or conductor merely by moving it in a strong magnetic field . . . thus was born the first rudimentary dynamo'. These quotations are from the distinguished pen of J. A. Fleming.* This principle is the basis of nearly all electrical machinery, apart from devices for communications, and the electricity supply industry depends on it.

'Discovery' is not strictly the right word in the second case, because oil, the mineral occurring in nature, had been observed since ancient

* J. A. Fleming, *Fifty Years of Electricity*, The Memoirs of an Electrical Engineer. London, The Wireless Press Ltd, 1921.

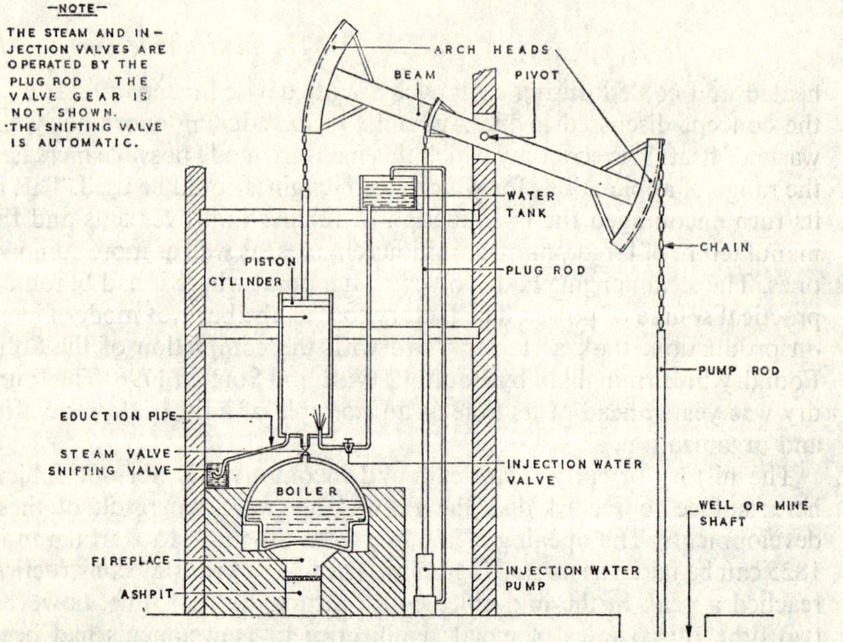

Fig. 2 The Newcomen engine. Injection valve is open, water condenses steam in cylinder and piston descends. Injection valve closes, steam valve opens. Steam enters cylinder, piston rises helped by weight of mine pump rods. Condensed water runs away through snifting valve. When piston is at top of cylinder steam valve closes, injection opens and cycle repeats.

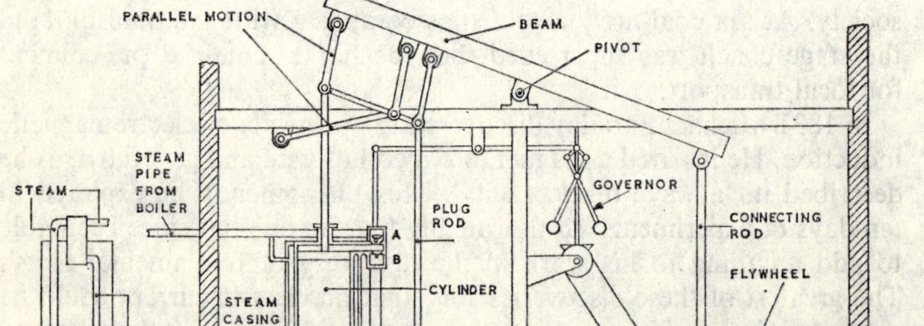

Fig. 3 The Watt rotative engine. A and C are steam valves. B and D are exhaust valves. A opens to admit steam to top of piston, D opens to exhaust steam below it to condenser. Piston descends. B and C open, steam is admitted below piston, which rises, steam above it passing to condenser. Cycle repeats. (Figures 2 and 3 are not to scale. Source: W. K. V. Gale.)

times. Modern usage of oil as a fuel started, however, in a modest way on the American continent, with the exploitation of the Pennsylvania oilfields in the U.S.A. by Drake in 1859, and by similar development on a smaller scale at Oil Springs and Petrolia in Ontario, Canada. The first oil well on the American Continent was drilled in 1858 by James M. Williams at Oil Springs. At that time the crude oil was refined to yield kerosene (paraffin), or 'burning oil' as it was called, which was used in lamps for illumination. Town's gas had been introduced by Murdoch (who had installed gas lighting in his offices and home at Redruth in Cornwall in 1792) so there was no great demand for kerosene, in terms of the quantities which would have justified expanding the oilfields, and gasoline (petrol) was a dangerous and unsaleable waste product of the refining. The *Oil Springs Chronicle* was certainly far-sighted and years ahead of its time when it published the comment in 1862: 'Petroleum will necessarily supersede alcohol (as a fuel). We may look for its adoption as a fuel for the generation of steam in our ocean steamers where economy in bulk and weight are desired . . . even crude oil is sought with eagerness in many workshops in England for lubrication . . . When mixed with fat or resin . . . it is more generally applicable . . . paraffine can be extracted for candles.' This comment could not foresee the development of the internal combustion engine, which, depending on oil, and not forgetting rubber tyres, ushered in the next great surge of development. 'Petroleum' in fact became a vastly important fuel, but the writer in the *Oil Springs Chronicle* would have been interested to learn that it was superseded in its turn for some special applications when German aeronautical engineers used alcohol as a fuel for aircraft rocket motors in the closing years of the Second World War. These two discoveries, then, initiated the chains of events which brought in the age of electricity generation, on the one hand, and the internal combustion engine, and hence the motor car and the aeroplane, on the other. Steam still has its place, and so does coal. But in some respects they in their turn have been superseded.

There is one more recent achievement of highest significance in this history: it is the construction and operation in Chicago of the first atomic reactor, known as an atomic pile, by Enrico Fermi in 1942. A chain reaction was initiated and sustained, thus taking the first major practical step in making an atomic bomb, and so opening the road to the utilization of nuclear energy.

How engineers led the scientists

The unit of energy in the SI system, the joule, is named after J. P. Joule of Manchester, who carried out experiments, as recorded in every textbook of physics, to establish the mechanical equivalent of heat and so put the first law of thermodynamics on an impeccable academic footing. This he did in about 1850. His work was notable for its display of experimental skill, but also because it added no new knowledge, other than an accurate determination of the mechanical equivalent, of course, to science or engineering. By then steam engines had been in use for over a century, and Carnot's penetrating analysis (*see* Appendix) had been published thirty years before. Carnot's work clearly depended on the notion of a mechanical equivalent of heat. To engineers who designed and built steam engines this concept was well known probably from the time of Newcomen, but perhaps in a more practical form and less explicit scientifically. To an engineer the 'heat rate' is an important measure of the economy of an engine. It measures not the scientific equivalent between heat and work, but the practically attainable one, that is the quantity of useful work that an engine will do for the consumption of a given quantity of heat or for the price of a given quantity of fuel – and that is the acid test in practical terms. Improvement of the heat rate continues to be an important aim, since cheaper useful energy comes with it in most cases. The range of temperature over which an engine works was also known to be an important factor in its economy before Carnot established the fact analytically.

These facts illustrate the unusual historical background of thermodynamics. Science investigates the natural universe; in this case it was led to investigate, some would say codify, the properties of an artificial device, the steam engine. Yet thermodynamics, and especially the second law, have influenced scientific concepts of the physical world fundamentally, and give understanding of many physical phenomena not remotely connected with steam engines. I place some emphasis on this point because engineering, or technology as it is usually called today, is sometimes considered to be a routine, mechanical exercise in applying scientific principles and methods, and, by inference, to be intellectually inferior. A more correct view is that engineering is a practical art; it borrows scientific theory if that theory assists in attaining useful ends, but engineering will attain those ends in any event, by some mixture of intuition, practical experience, and lastly theory, 'applied' or 'pure', and the borrowing is a part, sometimes insignificant, sometimes dominant, of that creative process. In the case of thermodynamics, science was the borrower. The practical achievements of engine-builders forced the subject on the attention of theoreticians, and presented to them profound philosophical vistas. The development of heat engines retains this spirit today, and to me it seems that those who design heat engines are undervalued intellectually by society at large. In the same way, the job of diagnosing a fault in a car engine involves a nicety of skill, experience and deduction, and this art too, although on a different level, tends to be undervalued. Improvement in the complex, practical art of building and operating heat engines has been one of the central achievements which have made possible the great burgeoning of energy usage which is the theme of this book, and the backbone of industrial society. Each of these improvements was gained only with perseverance, patience, knowledge and experience, and sometimes a touch of genius.

Energy on tap

One further line of evolution needs to be traced out a little in order to round off this historical résumé of the birth and salient points in the growth of energy usage. It will be clear that the heat engine – steam or internal combustion – has been the vital device. Then a store of fuel is kept, and burned when energy is needed. But there is one other concept of energy supply that has attracted many keen minds and practical talents – that of having energy on tap like water. This we have today with electricity, but a glance at history is instructive. The first attempt of this kind was William Murdock's* idea of piping gas to supply a town. Although he used gas in 1792 to illuminate his own house, the significant date is 1805, when a contract, the first of its kind, to illuminate a cotton mill in Salford was undertaken by Boulton, Watt and Sons. Gas is very easily burned, straight from the pipe, and it was useful for one high-grade use of energy which is vital to a town, namely to give artificial light. Murdock's suggestion that gas could be supplied in this way and used to illuminate streets (which had never been done before) caused vociferous opposition. The whole idea was considered to be eccentric, unsound and dangerous, and doctors warned that illuminated streets would encourage people to go out in the damp night air and so take chills and become ill.

Since steam was the universal medium for utilizing energy, it might have been expected that steam mains also would have been considered in a similar manner. It must be remembered, however, that steam was generally used only at low pressures, until well into the nineteenth century. The advantages of raising the upper temperature at which an engine worked, and increasing the energy content of the steam at that temperature still further by increasing its pressure, were well understood, and Richard Trevithick (1804) paved the way to the locomotive engine by using steam at pressures well above atmospheric. But in practice, risks of explosions in boilers, of leaks and heat losses were

* Born Murdoch in Ayrshire, died Murdock in Birmingham.

otherwise considered to be too great. The use of higher pressures for steam was one of the great controversies of those early days. That being so, the use of steam mains for any but local purposes was ruled out. It was not until the mid-twentieth century that steam mains for district heating became common in some countries. The prototype is a scheme at Mannheim in Germany, dating from 1936. District heating using hot-water mains is, however, much older, the first scheme at Lockport, New York dating from 1877.

The term 'hydraulic' is used in several rather different senses these days. Sometimes it is loosely left untranslated from French usage, to refer to hydro-electric power. It can refer to the study of rivers, and of aqueous flow. It is also commonly applied to a multitude of devices from an aircraft undercarriage to the operating mechanism of a bulldozer blade or an automatic transmission for a car, which transmit power or movement by the use of a fluid under pressure, usually oil. High pressures up to say, 1,000 lb/in^2 (Sorry!, 6·9 N/mm^2) are common today for these purposes, and some specialists are interested only in pressures more than five times greater than that figure.

Hydraulic power in the sense that is meant in this context is a more primitive version of these principles, using water under pressure. In this system, as typically used in the nineteenth century, the store of energy was a hydraulic accumulator – a tower-like cylinder full of water, made of cast iron, and a piston, an iron or steel ram, floating on the top of it and sealed with leather packings. Weights were placed on the piston to increase the pressure imparted to the water and the accumulator was replenished by pumping. The source of energy was thus a steam engine which worked the pumps. Pressure mains led from the accumulator, sometimes for miles, to points of use, where hydraulic cylinders utilized the pressure to do useful work.

The hydraulic accumulator was invented in 1851 and the system was widely used. Examples can still be found, mainly in specialized industrial applications, but one public utility – unique in the world, as far as I can ascertain – today supplies hydraulic power to customers within its area. The London Hydraulic Power Company (under an earlier name) was incorporated by an Act of Parliament in 1871, and started to supply power in 1883, when 7 miles of mains had been laid. Today it operates nearly 200 miles of hydraulic mains in central London. The company's year of peak output was 1927, but in the 1960s it operated at nearly half of its 1927 output. One example is Tower Bridge, where there is a hydraulic accumulator, topped up by one pumping engine. Hydraulic power starts the bascules moving quickly when the bridge is to be opened, and while other engines are put in motion.

By far the most successful of the 'on tap' methods has been electricity, which has superseded all the others for widespread and general application, although it may be supplemented by gas, steam or hot-water mains. The great variety of pipelines which we see in use today often carry fuels, and oil and gas pipelines are in some respects analogous to the high-voltage transmission of electricity, but pipelines are more in the realm of transport than of 'energy on tap'.

Electricity had been applied to illuminate lighthouses in England and France from 1857, but various developments in dynamos had to take place before industrial generation of electricity became a possibility. Thomas Edison was the principal instigator of the concept of a power station linked with public electricity supply, but others were active too. His power station at Pearl Street in New York, opened in 1882, is the significant prototype power station for all electricity supply in the same sense as Newcomen's engine. Like Newcomen, he was anticipated by similar developments – the power station he himself developed at Holborn Viaduct in London, and one at Godalming in Surrey* for example. To quote again from J. A. Fleming:

> Soon after 1878 a fresh stimulus was applied to invention in connection with dynamo-electric machinery. Mr Edison had done much more than attack the problem of producing a small-unit incandescent electric lamp for domestic electric lighting. He saw that the possibilities of such lighting were essentially dependent upon the public supply of electric current from a generating station in the same manner that gas or water is supplied. Also that this supply must be a constant voltage or pressure and the energy metered to the customer, and paid for by unit. He accordingly took large views and proceeded to think out all the details of such central station supply . . . Many practical difficulties had to be overcome before an uninterrupted supply could be given, and on many occasions it was necessary for Mr Edison to take off his coat and practically instruct the workmen in the laying of street mains or the running of dynamos in parallel at the supply station. At first it was considered quite a feat if a dynamo could be kept going for a day or so without some failure necessitating a stop for repairs. It was only by slow degrees that success in the construction necessary in the case of supply station dynamos was obtained. The author visited this New York station [Pearl Street] in September 1884 and was allowed to make a full and detailed report on its construction and commercial position to the Edison and Swan United Electric Light Company of London. Even at that date the

* See R. H. Parsons, *Early Days of the Power Station Industry.* Cambridge, Cambridge University Press, 39.

New York station had proved that electric supply could become a profitable industry. A similar Edison station was established at a little later date in Milan, operated with the same type of dynamo.

Electricity supply has progressed by leaps and bounds from that day to this, and shows no sign yet that it will not continue to do so for another century. The complete dominance of Britain in the first phase of the industrial revolution is clear, even from the brief historical account given here; the only exception in the period in our subject is the academic work of Carnot. But in the second phase the emergence of the U.S.A. is equally clear. Henry Ford's mass production of the motor car has a significance at this second stage similar to Boulton, Watt and Sons' Soho Foundry in the first. It is of interest that meeting the demand for illumination – a high-grade usage of energy but one requiring relatively small quantities – was the reason for introducing two 'energy on tap' systems, gas and electricity. Three 'energy on tap' systems – hydraulic power, electricity and district heating – were introduced within a short period of time.

There is a remarkable omission in this history, which is as significant as the discoveries themselves. Water on tap depends ultimately on a reservoir. How can energy be stored to supply any of the on-tap systems? The answer is that, broadly, it cannot. One can have a store of fuel ready to burn, a store of water in a reservoir at a high level in the form of potential energy, or maybe a hydraulic accumulator or a battery. That is about the limit of the possibilities, and the reservoir is the only method of storing directly any appreciable quantity of energy. Unless such a reservoir exists, energy must be created as it is used, by burning fuel in an engine or by generating electricity simultaneously with the demand for it.

Fuels and sources

To trace all the ramifications of this historical background as they have developed into the 1970s would be an encyclopaedic task. On the one hand there is the development of the coal industry, culminating today in highly mechanized underground work, or the use of gigantic earth-moving machinery for open-cast mining. Then there is that worldwide, uniquely organized and select group, the oil companies. Drilling for, and extracting, oil is a technology of enormous expertise and enterprise; and the construction of pipelines, terminals, tankers and refineries, equally so in each case.

The heat engine being the crucial device within our theme, then the theme itself could be given to that one subject: from reciprocating

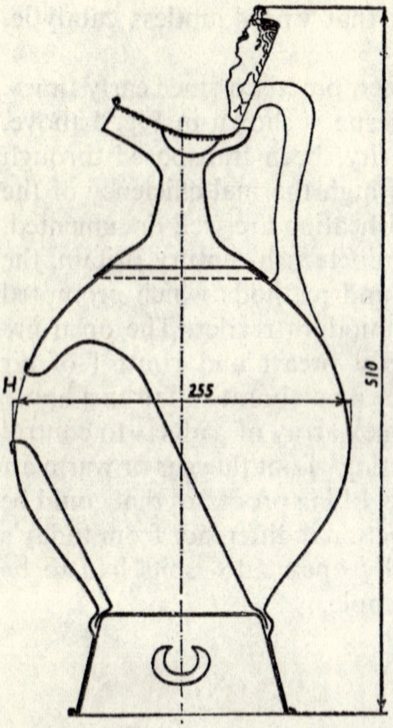

Fig. 4 The authespa. The tube H has been transposed through 90 deg to clarify the sketch. Dimensions are in millimetres.

engines to steam turbines, to automobile and aeronautical engines. Yet again, the transmission of energy, by trains of gears or by hydraulic methods, is a subject within our definition. So is the burning of fuels for rockets where the highest possible thrust per unit weight of fuel is the aim. As a practical limit, then, it is better to take for granted the world-wide sources of fuels, the winning of them, and their usage (but not the conversion of energy by mechanical devices) except when they influence the broad trends of the story of energy and its use.

Heat

The history of heat engines is well defined, but that of the use of heat alone is not. It is an anthropologist's puzzle to speculate on its origins. Cooking, and working metals with heat, are the most ancient of human arts. Dr Sugden, my lecturer in chemistry when I was a student, quoted an ancient description of how a Damascus sword was made, couched in extravagant terms, which he 'translated' into precise metallurgical instructions. For example: the colour of the setting sun; a precise instruction for temperature, since the colour of steel varies with temperature; plunging the sword through the body of a slave of good character – that case-hardened it with a nitride coating. The only thing he could not explain was the praying, but that was doubtless catalytic, he concluded.

Economy in the use of heat has also been practised since early times. A Roman cooking utensil called the authespa is shown in Fig. 4 above, in which the central tube H has, for clarity, been transposed through 90 degrees. Tests have demonstrated the high thermal efficiency of the authespa. Roman hot baths and central heating are well documented, but even in the heyday of cheap coal in nineteenth-century Britain, the widespread and effective use of devices and methods which promoted efficient combustion would surprise the modern reader. The open fire had an adjustable throat in the chimney breast and control of air entering below the fire. The range could be an elaborate affair in a house of any pretension, equipped with a complex array of gadgets to control rate of combustion, draught, and the ducting of hot flue gas or warm air to whatever was demanded by the many cooking processes that could be carried on simultaneously. The main functional difference from today's cookers was that the range was manually operated – spits had to be turned through pulleys by hand, for example.

4 Electra, the brilliant one

Apart from transport and the supply of heat, energy usage is becoming increasingly dominated by electricity. In all the countries of the world, once an electricity supply has been set up, demand begins to increase and continues to do so. In the less developed countries, demand for electricity now typically doubles in five to eight years; in the industrialized countries of the west, demand tends to grow more steadily. Usage of energy is increasing, but the proportion of that usage which electricity supplies also grows.

Exponential growth (i.e. a rate of growth proportional to size, such that doubling in size occurs at equal intervals of time) is a favourite topic of comment these days, largely because world population is growing exponentially. Perhaps there is a lesson to be learned from the fact that the electricity supply industry has been coping with exponential growth and doubling in size maybe every seven to ten years, for the whole of its life of nearly a century. Such a record presupposes considerable skill, foresight and perseverance. Given these qualities, it is a characteristic of technology that experience brings improvement in the design, the reliability and the economy of its devices. Electricity supply is a good example, for it has progressed steadily and quickly over the years.

Its starting point was Faraday's experiment. A generator – or alternator as it is also called because it makes alternating current (the term 'dynamo' seems to be obsolete now, except for Russian footballers) – has windings analogous to the coils in Faraday's apparatus, which are rotated so that they pass close to magnetic poles. As the windings cut the lines of force of the magnetic fields of the poles, an electromotive force is generated in them. Work has to be done to rotate the machine against these magnetic forces; it may be supplied by a heat engine (thermal-electric generation) or by a water turbine (hydro-electric generation). The water turbine will be run by a head of water. The heat engine will commonly be fuelled by oil or coal, or a nuclear source of heat, and most commonly will be a steam turbine. An auxiliary supply of power will be needed to energize the windings of the poles of the generator.

The usual arrangement of a group of machines for generating electricity thermally consists of a boiler – coal- or oil-fired – and a turbine and generator mounted on the same shaft and called a set in British parlance, a unit in American. The unit principle will most probably be followed for the boiler which will have a capacity matching the demands made on it by the turbine, and for other ancillary elements of plant, like the group of electrostatic precipitators which catch the dust from the boiler flues before the exhaust gases are discharged.

There is an advantage in large size, and with increasing experience the size of sets has increased. The size of individual sets is limited to about 10 to 15 per cent of the load on the network which the power station supplies. Operating difficulties arise if a generating set larger than that limit is suddenly taken, or thrown, off load. If the designers can overcome their problems, therefore, as the network grows so larger sets can be installed. One other advantage sought with increasing experience is the thermodynamic one of increasing the range of temperatures over which the turbine works, and increasing the total energy of the steam at the upper temperature by increasing also its pressure.

During the last two decades or so, progress in increasing the size and the thermodynamic conditions (with modest but worthwhile improvements in thermal efficiency) of turbine-generators has been rapid. Cheaper electrical power has thus become possible, and the big 'conventional' thermal sets have been setting the pace for other types of generation. The generation of steam by coal- or oil-fired boilers linked with these large turbine-alternators is thus today the basic method of electricity generation. It is standard practice to install a few sets of great size in a power station which will burn perhaps 20,000 tonnes of coal in a day, which demands constructional work on a vast scale to build it, and machinery and processes which are elaborate and of high performance for it to operate successfully. That is normality, and is taken for granted, but to give some insight into what is involved, Chapter 11 includes a description of one modern coal-fuelled power station.

Nationalization of the British electricity supply industry a quarter of a century ago presented an opportunity for standardization on a national basis for all new plant, and nominal capacities of 30 and 60 MW were selected as the standard sizes. Many of these sets were built, but in retrospect it is clear that they were too small. Demand grew fast, and they were superseded by sets of 100 and 200 MW. A second and much bolder attempt at standardization was made for the sets to be installed in the mid-sixties. The size chosen was 500 MW, and the sets were generally to be grouped in four, in power stations each of 2,000 MW, there being considerable standardization in the power station as well. Although a set of such size had never been built or operated anywhere in the world at that time, orders for 47 of them were placed.

There was enormous economic advantage to be gained from carrying through the programme for these gigantic sets, much of which is now being realized. On this occasion, however, the tendency was for the standardization programme to be too ambitious, and there were long delays in commissioning the 500 MW sets, partly because of their novelty and size. A 2,000 MW station cost, in the sixties, nearly £80m to build (i.e. £38 per kilowatt installed). The penalty for delay if one set of one of these stations was not completed and ready to go on load at the scheduled time was variously estimated to be from £10,000 to £20,000 per day. Consequently their construction was organized with great attention to maintaining progress as planned. Nevertheless, there were serious delays either of construction or operation, and a government committee (the Wilson committee, whose report was published in 1969) concluded that the large increase in the volume of plant ordered and the rapid advance of technology had overloaded the resources – of staff as well as of production – of the whole industry. All but two of these sets were in use by the summer of 1973, and their advantages are now being realized.

These power stations, needless to say, have already been superseded, a fate which overtakes the typical power station early in its life, maybe even before it has been built. In recent years the Central Electricity Generating Board (C.E.G.B.), the nationalized main British electricity generating organization (distribution and sales in Britain are handled separately by area boards) has had probably the world's largest programme of building and commissioning new thermal power stations, because it is national in size (Scotland has separate organizations) and has virtually no hydro-electric resources; in the U.S.A., the same task is divided regionally into various separate, privately owned organizations, the largest of which are similar in size to the C.E.G.B. In Russia, regional divisions, hydro-electric resources, etc., break down the national programme into separate parts.

Other thermal methods

When the load is not large enough to justify the installation of a steam-turbine-driven set, a set driven by a diesel engine is frequently selected. Such sets are relatively cheap to install, but relatively expensive to run. They are robust and reliable, and are the natural choice for undeveloped or isolated places, or where there is not much demand for electricity. At least that is the usual opinion; a different one is explained in Chapter 9. Gas turbines have been adapted to generate electricity. They are used as stand-by capacity, but primarily to meet the extreme peak loads that occur perhaps twice, for a total time of an hour or two, each working day. For these purposes they have the advantages of being quick and simple to start and run up to load, and like the diesel set but even more so, cheap to install but expensive to run. Their advantage is lost if they are operated on base load.

Water power

A water turbine with a generator coupled to its shaft can pick up 80 to 90 per cent of the energy of the water and convert it into electricity. The Pelton wheel is physically the simplest type of water turbine to understand. It needs a high head and is mostly used for heads over about 500 metres, up to the maximum of around 2,000 metres so far exploited. Jets of water are directed at the turbine wheel – the runner – and strike cup-shaped deflectors which absorb the kinetic energy of the jets as the runner rotates. For an intermediate range of heads, say 50 to 600 metres, the Francis turbine is likely to be used. Changes in pressure as well as in velocity take place as water flows across the blades of the Francis turbine's runner, which are shaped so that streamlined flow occurs (turbulence causes losses of energy). Physically, there is no easy way to visualize how the Francis turbine works, but it is an intermediate stage between the Pelton wheel and the low-head turbine – propeller or Kaplan machines. A propeller turbine is a ship's screw in reverse, taking energy from the water to drive the generator. In this mode of operation, the screw is sensitive to variations in head and load, so the Kaplan turbine has a servo-mechanism that adjusts the angles of the blades so that they can work close to peak efficiency over a range of heads (up to almost 100 m) and loads.

These are the three basic types, but there are several others, like the Deriaz turbine which is a reversible machine with adjustable blades that can be used both as a turbine and a pump, and the bulb turbine for low heads and tidal power. Improvements over the years have increased the size of water turbines and the ranges of heads over which the various types operate; these improvements have been appreciable, but not as great as with the corresponding evolution of the steam turbine. The vast growth in competence that has taken place in the development of hydro-electric power has been largely on the civil engineering side. In principle it is now possible to control and utilize every drop of water even in the largest river catchments of the world.

Electricity cannot be stored; water can. So many hydro-electric schemes have large reservoirs, sited as high in the catchment as possible. Considerable quantities of energy can be stored in these reservoirs. In the typical Alpine scheme, as the winter snow melts in May or June, it is impounded in a reservoir, maybe 2,000 metres above sea level. In the following autumn or winter, the reservoir is drawn down and power generated from its waters. In other cases the purpose of the reservoir may be to regulate the river so that an even flow is available for the hydro-electric power station over a year or a season.

Small reservoirs provide 'pondage' sufficient to operate the power station for an hour or two, and at the other extreme over-year storage is needed on some schemes, and on the Nile 'century storage'. Impounding may be necessary to create a head, rather than for storage. As the river flows towards the sea, it acquires a different character in its lower reaches, and if extensive regulation has not been effected upstream, a run-of-river scheme (i.e. one with no storage) is likely to be the best form of utilization that can be accomplished.

Water is the commodity that takes first place amongst the vital needs of living things; the control and use of rivers, therefore, has other purposes besides the generation of power. The large dam is the key structure in river control. Regulation of the river is essential before its resources can be managed, and for this the dam is needed, so it may serve purposes other than the generation of power, or power may be one of its several purposes, or its only purpose. Apart from the dam, a hydro-electric scheme may involve a great deal of tunnelling, and the power station itself may be underground. The control of flood waters is a critical aspect of many schemes, since water that is discharged uselessly has as much energy as that which drives the turbines, and its destructive capacity is immense. The most spectacular method of flood discharge is the ski-jump spillway.

Transmission

There is more to the success of electricity supply than the power station. High-voltage transmission of power is one of its cornerstones. Losses of energy in transmission are reduced if the voltage is increased; so if the problems of insulation and switching can be solved, larger quantities of power can be carried over longer distances. In a country like Sweden, this has been a critically important problem. Sweden has abundant, cheap resources of hydro-electric power in the far north of the country, but the centres of consumption – the cities – are hundreds of miles farther south. High-voltage transmission enabled the sources to be linked economically with the cities. There are other striking examples of this kind; what was revolutionary a few years ago is now accepted as conventional practice, and the transmission of large quantities of power over long distances is now practised in many countries. Voltages increased as progress was made, and an interesting trend emerged of using direct current at very high voltage for long-distance transmission.

Research is in progress to investigate the limits to which these developments may be pushed, and experimental rigs working at up to a million volts are being tested.

It has been announced that the main transmission links in Russia are, in the future, to be at 1·2 MV. The first experimental line at this voltage will be started in 1975. This very high voltage is expected to reduce losses by 20 per cent compared with a line operating at 0·5 MV.

Apart from the job of carrying power from place to place, the high voltage lines 'interconnect' the various power stations and centres of consumption of the electricity supply system. Alternative routes of supply give greater security against faults, but there are more advantages to interconnection than that, and it permits major economic gains. The margin of generating capacity held in reserve can be reduced, the most efficient power stations can be operated on base load and the size of generating sets and of power stations can be increased. These facts were realized rather earlier in Britain than other countries, and the British system has always been well served by its grid.

The technology of high-voltage transmission is an erudite sphere of study, dominated by mathematics. By contrast, the physical reality – a line of gaunt, steel-framed towers with about six conductors strung high in the air between the towers, which are slung from great insulators on cross-arms – is crude in appearance. I remember seeing a transmission line snaking its way up Stelvio Pass (between Italy and Austria), the towers standing improbably on craggy bits of mountain, which added visual high voltage to the scene. But a similar line passing through the modest scale and intimate greenery of rural Sussex is a visual intrusion. When several of these lines diverge from a switchyard, several acres in extent and full of scraggy poles and switchgear, not even the hardiest enthusiast for electricity can deny that it is an eyesore. There is little hope of an alternative, except in a few special cases. 'Undergrounding', as the electricity people say when they mean burying, is a very expensive process. Even though the losses, which appear as heat, are proportionally small, the capacity is great, and the quantities of heat that have to be dissipated are large; underground, special arrangements have to be made to do what the air does unaided, up aloft. Equally, insulation is much more complex.

The transmission line ends at the substation, where the voltage is reduced to an intermediate value which can be handled by the modest 'sub-transmission' stage, as it is sometimes called. This stage in its turn leads to the local substation, for transforming down again to the supply voltage, and thence through the main switches and meters of the buildings of the neighbourhood. These last links in the journey of electricity from the generator to the consumers' gadgets receive much less attention than the earlier ones, both in news and comment and in technical writings; they are nevertheless important, and their proper and efficient functioning is crucial to any system of electricity supply.

The cumbersome nature of electricity generation

Success and growth have been the themes of this description of electricity supply. As the years passed, the convenience and price of electricity encouraged its use. This in its turn depended on continued improvement in its technology – 'order of magnitude' improvements to match exponential growth. The record, and familiarity with accepted practice, tend to overshadow the fact that the accepted system is cumbersome and wasteful. There ought to be something much better, but there isn't.

This comment applies with much less force to hydro-electric power, a form of power generation that is fundamentally efficient and benign. Losses are small, and the source of energy is like a crop and is renewed each year by the cycle of nature. Theoretically a perfect solution would be possible if water were magnetic, and so had its own magnetic field, as mercury does; then the liquid could be directed as a jet, through coils of conductor, and Faraday's electromagnetic effect would occur directly, without the intermediate mechanical stage. The turbine would be unnecessary, and the generator fundamentally simplified. The theory of this method has been worked out with some elegance, but water is not like that so it can't be done. However, the same perfect approach theoretically to generating electricity has been toyed with for thermal power using hot, ionized gases (M.H.D. = magneto-hydrodynamic generation). Analytical deduction from scientific principle does not necessarily lead to a practical answer (more's the pity) so confirming the notion that engineering is a practical art, not low-grade science.

Whilst hydro-electric power is the ideal method of generation from the point of view of conservation of resources, and ecologically in other respects too, its scope is limited to the amounts, and its generation to places, determined by nature. There are few countries fortunate enough to have hydro-electric resources which are sufficient to meet all their needs for electrical energy. There are also abundant resources in remote places; the greatly increased range of use of transmission lines at very high voltages has brought major new sources, e.g. in Quebec and Labrador, into exploitation, but others are far too remote from any centre of electrical load. For example the largest hydro-electric site in the world is near the Tibet–Assam border, where the Brahmaputra River, descending from the Himalayas, flows east, and then, at the eastern limit of its course, turns in a great loop towards the west. By driving a tunnel across the loop, between 17 and 38 GW of power could be generated continuously (compared with 6 GW installed, but not operated continuously, at the world's largest hydro-electric power station at Krasnoyarsk on the Yenisei river, Siberia) but the nearest centre of demand would be Calcutta which is 1,000 kilometres away.

Hot water that is not very hot

'Conventional' thermal generation can, at the best standard of modern practice, achieve a thermal efficiency of about 35 per cent; over a whole network, the average thermal efficiency is much less (e.g. for the C.E.G.B. in 1972 29 per cent*). So about two-thirds of the energy of the fuel is thrown away in the initial stage of the process, before the inevitable inefficiencies and losses of later stages – transmission and distribution and losses in actual consumption – occur.

The heat rejected by a power station is equivalent to the energy of burning half to two-thirds of its coal. This heat has to be dissipated, either into the atmosphere through cooling towers or into the waters of a river or the sea. The outlet and inlet temperatures of the circulating water which cools the condensers differ by 7 to 8 deg.C. The outlet temperature is kept as low as possible, say 30 deg.C maximum.

A fortune awaits the man who can find a use for oceans of hot water that is not very hot. Many attempts have been made. One successful approach to this problem is known as 'heat-electric' generation, meaning that heat and electricity are both produced. The use of a back-pressure turbine or a pass-out turbine is entailed. Instead of using every joule of energy of the steam down to the lower limit of temperature and pressure of the cycle, a small part of its useful energy is retained in the steam exhausted from the turbine (as implied by the terminology) which is then a useful source of heat in its own right. In special cases where steam is needed in industrial processes this method is highly advantageous. Otherwise, the steam is used for district heating. Steam mains can be laid to supply heat for the buildings in central city areas or other densely built places such as blocks of flats or industrial premises. The thermal efficiency of this combined system, with both heat and electricity on load, is around 80 per cent. One difficulty is that the quantity of steam tends to outrun the possibilities of using it, long before the electrical demand is met. Another difficulty is balancing the heat and power loads; at peak demand for electricity the demand for heat may not be a maximum. Day-to-day fluctuations can be evened out by building a

* The C.E.G.B. expects an improvement of 0·7 to 0·8 per cent in the overall average annual thermal efficiency of its system in 1972–73, to reach 29·7 per cent, a rise of 0·1 per cent giving the C.E.G.B. a saving in the cost of its fuel of £2 million.

heat reservoir – a tower full of hot water is the form it has taken. Seasonal fluctuations cannot be coped with at full efficiency, and if the demand for heat disappears and that for electricity remains, there has been a penalty in operating such turbines at a lower efficiency than if they had been designed for supplying electricity only. British electricity planners are confirmed in their view that there is no economic advantage in a mixed heat–electric supply system. The high-grade product, electrical energy, is the one in which they are interested, and which they consider should be supplied at the lowest possible price. This is best achieved, they consider, by exploiting the advantages of large generating sets, operated for one purpose and at the maximum efficiency for that one purpose, i.e. with the production of electrical energy maximized. Such sets are in themselves highly complicated mechanisms, without adding further complexity by attempting to get two products.

Most electricity authorities throughout the world have concurred in this view, with one outstanding exception, namely Russia. Heat–electric generation is widely practised in that country. About half the capacity of recent installations of new generating plant is in heat–electric sets, some of them unprecedented in size (250 MW electrical capacity) for this combined duty. It would appear that the Russian climate is an important factor in this policy. Perhaps another factor is that a different economic evaluation has been made. One assumes that the overall benefit to the community of efficient heat together with electricity supply is greater than a slightly cheaper electricity supply and an uneven efficiency in the supply of domestic heat. Domestic heating is notorious for its inefficiency and waste of fuel; a competent and efficient method replaces it in this case. In my view this is a farsighted policy. Whether there is economic gain cannot be stated firmly without an exhaustive comparison, but clearly the Russians consider that there is. In recent years there has been much more attention given to this method in some other countries. At Orebro in Sweden, three-quarters of the population of 92,000 are connected to the district heating, giving a heat load of about 450 MW. A turbine set of 106 MW electrical capacity now supplies about 200 MW of the heat load.

There must surely be favourable sites (i.e. for a power station plus a district heating demand) in Britain where a modern version of this system of reasonable scale could be installed. The advantages of a highly efficient supply of heat from the district heating system would not be as great as in Russia, but should be sufficient to justify the scheme. The first scheme of a new kind is almost always the most expensive, and experience brings improvement, and also more accurate estimation of costs and benefits. When such experience is entirely lacking (with minor exceptions) it is difficult to have any confidence that the policy of electricity alone is right for 100 per cent of new installations. There should be room for at least a share of this type of installation in Britain, maybe considering electricity as a byproduct of the plant, and heat its main product. Neglect of combined heat–electric generation, however, is part of a tradition of backwardness in Britain in district heating generally.

Distilling sea water

One other useful application of the waste heat of power generation is for the distillation of sea water, where fresh water is scarce or absent. Power stations exhausting heat to multiple-stage flash evaporators have been constructed in arid regions. It could well be that distillation of sea water will be needed in the eastern part of Britain perhaps by the end of the century. The best method would probably be by building a nuclear power station and utilizing its waste heat. There is no problem of balancing the heat and power loads in this application because the distilled water can be stored.

Fish-farming and hot houses

Other methods of using the waste heat of power stations seem to be more in the realm of research or imagination than of practice. Fish-farming has been practised in the local reaches of estuaries where the water is warmer than in nature because of the proximity of a power station's discharge, but the fish farm, if successful, is a *pis aller* and could never begin to justify the waste of energy. One other use of a similar nature commonly advocated is to site glasshouses close to the power station and heat them from the power station's waste heat. There seem to be a number of important practical objections to the idea, but it has been done in Rumania.

Peak loads

The perennial bugbear of electricity is, however, the peak load. There must be sufficient generating capacity in the system to meet the peak demand, with a certain reserve for breakdowns and plant out of service. This peak requirement dominates the planning of many electricity utilities, and numerous policies and devices are deployed to reduce the cost of meeting the peaks and to ensure that the most economical generating plants are utilized on base load.

Nuclear plants must be on base load. Run-of-river hydro-electric plants must be operated when the water is there. The most modern coal- or oil-fired stations also are put on base load; those less efficient come into service in the periods of middle demand. The extremes of peaking are met by gas-turbine-driven sets, or by drawing down the reservoirs of hydro-electric schemes. One other method has proved to be expedient, and that is pumped storage. It pays to build a reservoir which has to be filled by pumping – energy being supplied from the grid to the pumps during the night or when the demand is low – and then to use the water stored in this way to generate power during the peak hours. The efficiency of the dual operation of pumping and generating will be about 60 per cent or so but this elaborate and costly procedure is still often the cheapest choice for meeting the peak load, particularly if there is a nuclear station nearby which is more than adequate in capacity to meet the base load during the night.

A machine like a gas turbine which can be started and run up to load very rapidly is a valuable alternative method. The only other approach is to encourage more even consumption by the customers, by adjusting the tariffs which they have to pay. Off-peak tariffs, for example, encourage the use of devices like night-storage heaters, and expensive peak-hour tariffs discourage the use of plant at the peak hours. There is room for considerable advance in the way electricity utilities handle this problem.

Operating the system

Whilst the peak load dominates most supply systems, it has to be seen in the context of typical day-to-day operation. In Britain, this entails a variation in demand of about 6 to 1 between the winter peak load and summer base load. Such a sharp variation is against the interests of economy. The graphs (Figs. 5 and 6) show typical daily and seasonal variations. The important facts are that plant has to be brought on load at a rate of 20 GW in a few hours (say 150 MW per hour for individual plants or stations) and the utilization (i.e. the number of hours a year on load) of the plant which meets the middle region of demand is too low. The contribution of nuclear plant is limited to the size of the summer base load, until it becomes economic for nuclear plants to be built which can operate on a two-shift basis.

Tariff reforms seem to be urgently needed, but have to be implemented with care. The off-peak tariff for storage heaters actually produced a new, afternoon peak in the system.

It is instructive to look for a place for combined heat–electric generation within the conditions of these load curves; it would seem to lie in the electrical capacity by which the winter base load exceeds the summer base load, i.e. the demand for base-load plant during the winter only. The capacity from the curves would be a maximum of around 13 GW.

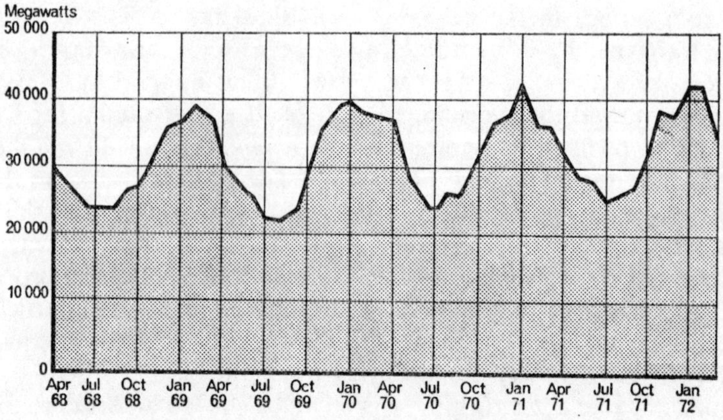

Fig. 5 Maximum demand for electricity on the C.E.G.B.'s grid (i.e. public supply for England and Wales) April 1968 to March 1972.

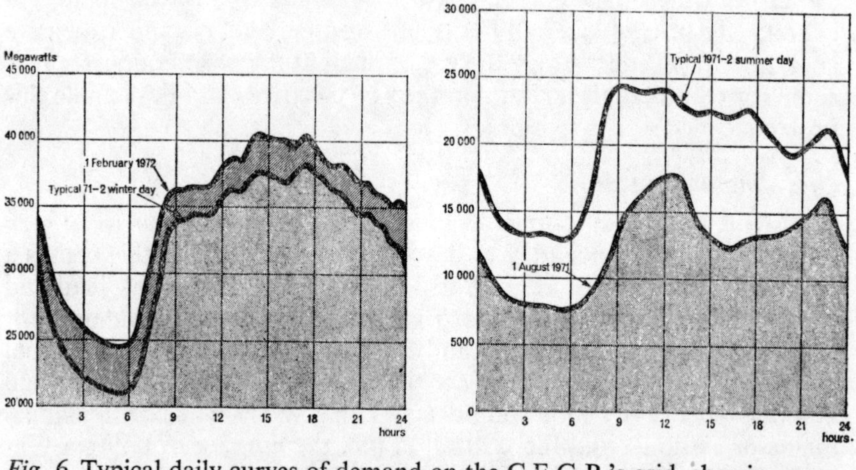

Fig. 6 Typical daily curves of demand on the C.E.G.B.'s grid, showing summer and winter 1971–2, and the days of maximum and minimum demand.

5 Mass into energy

For practical purposes, exploitation of all sources of energy except nuclear energy happily rests on a foundation of nineteenth century physics; that is, on the assumption that the atoms of an element are indestructible and immutable. Modern physics gives a newer interpretation in its description and explanation of the behaviour of the elemental particles of which all atoms are composed. The central fact which applies to our subject is that enormous amounts of energy are liberated in some processes of atomic mutation. The source of this energy is mass. Mass is destroyed and energy created according to Einstein's equation $e = mc^2$ (e is energy, m is mass, and c is the velocity of light).

The motive for utilizing this source of energy was destruction. The opportunity for utilizing it arose from war and from the discovery of nuclear fission. Adapting this work to the purposes of peace was as great a challenge as making the bomb, and offered hope of inexhaustible energy, an eternal benefit in place of the ultimate in destruction, followed by the certainty of radioactive poisoning for any lingering survivors. These were the prospects thrust upon the world as deductions from the abstruse thoughts of a handful of physicists. The dilemma remains; it colours human affairs, and in particular the story of energy.

To understand and control matter itself is a power greater than the medieval dream of the philosopher's stone. The use to which that power will be put tests humanity pitilessly, as if the human race were playing Faust. Nuclear energy may be man's greatest blessing, or it may destroy his existence and his culture. Just as success in medicine gave rise to staggering problems in containing the world's population, so success in harnessing energy presents staggering political ones, in preventing nuclear warfare. These are the stakes of modern life.

Fission, fusion and the ultimate explosion

The starting point was the discovery that some elements, such as radium, are radioactive. Radioactivity consists of spontaneous emissions from the atomic nucleus of three kinds: namely neutrons, which are particles without electrical charge and having a mass about equal to that of the hydrogen atom (that is, about one atomic mass unit); electrons, of unit atomic negative electrical charge, but negligible mass compared with a neutron; and short-wave radiation (x-rays).

Uranium is radioactive. It occurs naturally as a mixture of two isotopes of atomic masses 238 and 235, in proportions by weight of 140:1 respectively. The isotope U-235 has another property which was discovered in 1939. When an atom of U-235 is struck by a neutron, it splits into two. This splitting is called fission, and when it occurs a small part of the mass of the atom is transmuted into energy. The chemistry of the split pieces is rather complex, but they settle down – if such a homely phrase is permitted in this context – into atoms of other elements. In addition, two or three neutrons are ejected; the energy created by fission is in the form of the velocity of these neutrons.

If at least one of these neutrons collides with another atom of U-235, then a chain reaction is initiated, and enormous quantities of energy are generated while the reaction can be sustained. This chain reaction occurs spontaneously in a mass of U-235 larger than a certain minimum, the critical mass. The critical mass is 'some tens of pounds'.

There is one other route – discovered in 1940 – to a chain reaction of this kind. If U-238 is bombarded with neutrons, plutonium, a radioactive element not found in nature, is formed; it has an isotope Pu-239 that also has the property of spontaneous fission; its critical mass is apparently about 16 kilogrammes.

To achieve a nuclear explosion during the Second World War, then, depended on two lines of development. First, on separating the two isotopes of uranium, to concentrate the fissile U-235; for detonation, bringing together two masses of U-235 each a little below the critical mass, and building a container in which this could be done that was strong enough to hold together until the chain reaction had built up. Second, an essentially similar method based on Pu-239.

Isotopes are chemically identical, and there were only two possibilities for separating the two isotopes of uranium. One was diffusion, through porous membranes, of a gaseous compound of uranium. The rate of diffusion is related to the density of the gas. Lighter gases diffuse faster, so molecules containing the lighter isotope tend to pass more quickly through the membranes. The difference in density is extremely small, so the process is immensely complex and uses large amounts of electricity to pump the gas through the membranes. Uranium hexafluoride is a gas commonly used in diffusion plants. Secondly, a similar effect could be obtained by passing electrically charged gas through magnetic fields.

The magnetic process was first successful in making U-235 that was 85 per cent pure, and so was concentrated enough for the bomb. However, plutonium made in a reactor in which the heat was dissipated by cooling water could be separated chemically without great difficulty, and of the first three bombs two were made with plutonium.

The hydrogen bomb, which came later as the next major stage of development, has immensely greater destructive power than the atomic bomb, a thousand or so times greater. This even greater effulgence of energy comes from nuclear fusion. At a temperature of hundreds of millions of degrees, atoms of the lighter elements such as hydrogen and its isotopes fuse together to form heavier elements, such as helium. This is the reaction which takes place on the sun and is the sun's source of energy. It was achieved terrestrially by using an atomic bomb as the 'detonator' to reach the very high temperatures needed to initiate the fusion reaction. Explosions equivalent to 50 to 100 megatons of chemical high explosive appear to be the upper limit of size of hydrogen-bomb explosions.

About one-thousandth of the mass of the reacting particles is transmuted into energy in the fusion reaction. If the mass-energy transformation could be taken to its theoretical limit an explosion a thousand times more powerful than the hydrogen bomb would result. Anti-matter (i.e. particles of identical mass but opposite charge to the fundamental particles of matter) and matter would react to annihilate the total mass and generate preposterous quantities of energy. Quite a small bomb would blow the world apart and destroy everything on it. Table 11 lists the amounts of energy in this hypothetical case, and for the fission and fusion bombs.

A development at the other end of the scale of size was announced by the U.S. Government in May 1973. It had been impossible previously to make a small atomic bomb, and an explosion equal to about one kilotonne of high explosive was perhaps the extreme lower limit of size. But now tactical – as opposed to strategic – nuclear weapons will be in use in a few years from the time of the announcement, equivalent to about 50 tonnes of T.N.T. They will be devised on the one hand to maximize explosive blast, and to be without radioactive after-effects, or on the other to have minimal blast, but to 'explode' with lethal radiation of short duration. The nuclear holocaust is thus now to be available in a compact, almost domestic, size, not followed by indiscriminate radioactive poisoning. At the lower end of its range it is now quite indistinguishable, and just as sordid, as 'conventional' warfare.

Controlling nuclear energy

Prospects for controlling what is potentially an atomic bomb and persuading it to release its energy at a useful rate would seem to be bleak indeed, but within the wartime veil of secrecy a fundamental step in that direction had been taken a few years before the first atomic explosion, as part of the research in developing the bomb, when Fermi built his reactor.

A reactor demonstrating controlled nuclear fission is an invaluable tool of research for the physicist, but is a long step from a power station. That step was taken first in Britain when Calder Hall power station was opened in Cumberland in 1956. Critics of Calder Hall said that since it produced plutonium, which at that time was useful for weapons, it was really a military expedient; this view does less than justice to the idealism of its promoters and designers, who undoubtedly saw it as an important prototype, the purpose of which was to prove that a new era of abundant energy had arrived.

Calder Hall (Plate 1) has proved to be a remarkably reliable power station and functionally it has fulfilled everything that its designers intended. It is the prototype of the first generation of British nuclear power stations, known as magnox stations, which in 1972 generated about 10 per cent of Britain's electricity. They are now obsolete because their capital cost is too high. Subsequent events modified the idealistic optimism of the early period; one reason for this was the rapid progress made by conventional thermal generation, and both nuclear and hydro-electric power have had to face much stiffer competition. Only one or two magnox stations were built overseas. Nevertheless the magnox stations, as a group the first to make a sizeable contribution to electricity generation, have done well.

By 1971 the total electrical capacity of the eleven British magnox stations of 5 GW was in service, all of it on base load duty. These power stations have proved to be hard-working and reliable. They have been sited so that the country is nearly self-sufficient regionally in power (previously there had been heavy exports of power from the Midlands where the resources of coal lie) and their construction has fostered a strong nuclear industry and a tradition in safety of 'considering every possibility with the maximum pessimism' which has been widely rewarding since it has led to acceptance of this method of power generation by society at large.

How the nuclear power station works

For all the nuclear power stations yet put to work, as distinct from research projects, the fundamental principles are these. The reactor must be put inside a containing vessel which imprisons radiation and is strong enough not to burst if the reactor goes wrong. A mass of uranium, generally as oxide, itself enclosed in containers – cans or tubes – so that the products of fission can be removed, sustains the chain reaction which is the source of energy. There must also be a moderator; controls which can impede or stop the reaction, and a cooling medium which extracts the heat generated by the reaction.

Two or three neutrons are released when an atom of U-235 is split, and they are expelled at high velocity. Fission is induced more readily in U-235 if it is bombarded with neutrons of relatively low velocity. The moderator slows down the neutrons (by collision with the molecules of the moderator, which thus gets hot) to this optimum velocity, and it forms a sort of matrix into which the uranium is fitted in its containers, like honey in a beehive, so that the neutrons pass through the moderator and back again into the uranium. An important property of a moderator is that it must not absorb neutrons; by contrast, control rods when inserted into the core stop the reaction by absorbing neutrons.

A moderated reactor has a power output, volume for volume, which is of about the same intensity as the cylinder of an internal combustion engine, and it is the bulk of the moderator within the core which gives it this character, in contrast to the fast reactor which is described later.

The cooling medium has a simple, if vital function; it is either steam or circulates to heat exchangers where it transfers its heat to steam. From that point on there is no further novelty, simply the paraphernalia of turbines, generators and their auxiliaries just as for conventional thermal generation.

There is as much room for choice in applying these principles as there is in the choice of physical form for a heat engine, given that it requires a hot source, a working agent and a sink. So it is not surprising that at the start of the nuclear era an almost endless stream of variants on this theme was experimented with, in the hope that economic success would be the prize for the right choice.

Natural uranium, not enriched with fissile U-235, can sustain a chain reaction in an appropriately designed core. That fact was the basis of the Calder Hall type of reactor. The moderator is graphite of very high purity, and the uranium is contained in cans which do not absorb neutrons. The material selected for the cans was magnesium alloy, from which came the name magnox for these reactors. Their cooling medium is carbon dioxide gas, which has good safety properties, being inert and not supporting combustion.

The exhaustive efforts made in the U.S.A. to harness nuclear energy can be conveniently brought into the narrative here. It was American policy in the early days to investigate the possibilities systematically, so that one of the research reactors would pay off. The hope was realized and the U.S.A. light-water reactor has the edge on its rivals at present, and is competitive with other methods of power generation in some circumstances. Two American firms have developed and marketed light-water reactors, which have been installed extensively – 160 reactors at the beginning of 1973, of which 29 are operating, 55 being built and 76 ordered and being designed, with a total electrical installed capacity of 142 GW. This type has considerable operating experience to its credit and it is the only nuclear system that can make that claim just now, because the magnox system has been withdrawn and its successor, the advanced gas-cooled reactor (A.G.R.), is not yet in service and so has no operational experience.

There are two basic types of light-water reactor. In both of them water is both the moderator and the cooling medium. In the boiling-water reactor, to simplify a little, steam is generated in the reactor and is supplied to the turbines; in the pressurized-water reactor, pressure prevents the water boiling, and it is circulated to heat exchangers. The nuclear fuel consists of small pellets, which are contained in long tubes made of zirconium alloy, complicated assemblies of these tubes forming the core. An essential difference from the magnox reactor is that the fuel is enriched, having 2 to 3 per cent of U-235, as against 0·7 per cent in natural uranium. More than 100 tonnes of uranium fuel (oxide) in thousands of fuel rods, each about 4 metres long, will be loaded into the core of a light-water reactor of, say, about 1,000 MW electrical output. Such a reactor could, in theory, run for about eight years without re-fuelling, giving an output equivalent to about 20 million tonnes of coal.

The light-water reactor was given a flying start by Admiral Rickover's naval reactor programme, and it is reactors of this type which power the fleet of U.S. nuclear-powered (and nuclear-armed) submarines, the first of which, the *Nautilus*, was launched in 1954. Nuclear power has since been extensively used for naval vessels, and *Nautilus* was a prototype at sea as important as Calder Hall on land.

Canada also contributed to the early development of nuclear power, and a Canadian reactor system has been installed in several power stations. Many other systems have been dropped, and technology is moving on – the advanced gas-cooled reactor for major British power stations now about to go on load, and the American high-temperature gas-cooled reactors and the steam-generating heavy-water reactor are examples of the types of reactors which will be installed in power stations in the next few years.

In summing up the progress of this movement, Lord Hinton* said,

> There has been a certain amount of promotional talk about nuclear power and a tendency to be less than wise in rushing into it. In the

* Graham Clark Lecture 1971. Institution of Civil Engineers, London.

western world, only three countries, the U.S.A., the U.K. and Canada, have played an important part in pioneering nuclear plant design, and I doubt whether, overall, any one of them has as yet made any money out of it when all non-military expenditure is taken into account. The philosophy of design has been different on the two sides of the Atlantic . . . The British progressed to the present A.G.R. which achieves the same initial steam temperatures and pressures that are currently used in fossil-fired plant. The U.S.A., on the other hand, followed their initial practice of using water cooling and this has enabled them to build reactors that are smaller and more compact than gas-cooled reactors but has prevented them from achieving high steam pressures and temperatures . . . there is probably little to choose between the British gas-cooled and the American light-water reactors, but skilful and energetic salesmanship has enabled the Americans to dominate world markets.

The breeder principle

Two reactions take place in the breeder reactor. When a neutron is captured by an atom of U-238, an atom of the isotope of plutonium Pu-239 is formed. When a neutron is captured by an atom of Pu-239 that atom reacts by fission and creates more than three neutrons per atom; overall, one neutron sustains the chain reaction, one is captured by U-238 to yield another plutonium atom, and the remainder make more plutonium, or are lost or dissipated. The neutrons must be of high energy for the second reaction to be sustained. There is thus no moderator, and the neutrons each have an energy of around one million electron volts (their initial energy) compared with that of about a hundred electron volts of each moderated neutron of the magnox reactor. The core is as compact as possible to reduce slowing down and parasitic capture of neutrons.*

Breeder reactors

The next main objective of nuclear power is to make more effective and hence cheaper use of the energy liberated by fission, and to arrange the reaction so that the products of fission are themselves radioactive, and can, in their turn, be utilized in a reactor – the reactor breeds its own fuel. The fast breeder reactors now being developed will give at least 50 times better utilization of uranium than existing reactors. The prospect is thus presented of a cheap method of generating power, relying on abundant reserves of uranium which would be too costly to exploit with

* Acknowledgment is made to 'Nuclear power – the future', by T. N. Marsham and R. S. Pease (paper at British Association, Leicester, 1972).

the slow reactors of present-day types of nuclear power stations. The fast reactors have a further advantage in that they can be effectively integrated into the management of a nuclear-fuelled economy to permit savings such as obviating the need for some enrichment plants for uranium, in addition to their own better utilization of fuel. The advantages which they offer are thus very considerable, particularly in the long term.

Physically, they are much more compact, almost to the point where the reactor can be called a watered-down bomb. A fast reactor for power generations with an output of 1,000 MW of electrical energy would have a core typically a cylinder, 3 metres in diameter by 1·5 metres high (i.e. about 10 cubic metres in volume), supplying heat at a slightly greater rate than two of the four 'unit' boilers of a conventional 2,000 MW thermal power station which are each 'the size of Canterbury Cathedral' (Chapter 11).

The diagram Fig. 7 and Plate 3 illustrate the principle of the prototype fast reactor (P.F.R.) now coming into use at Dounreay in Scotland. Its core is cooled by circulating liquid sodium, and its maximum rating – that is, the maximum rate at which heat is generated in unit volume of the core – is 425 MW/m^3. The P.F.R. has a thermal capacity of 600 MW, and an installed electrical capacity of 250 MW.

A fast reactor has been operated at Dounreay since 1959 in order to reach the stage where the larger, prototype fast reactor could be designed and built. The engineering problems – well, it is unfortunate if an author overdoes the superlative adjectives, but maybe I can sum it up by saying, contentiously, that most engineers surely pray they will never meet anything like it. The intensity with which the heat is generated, plus the necessity for complete safety in operation, are the crux of the matter. For example the fuel in the core is contained in fabricated fuel pins, which are the most highly stressed pieces in the reactor; they have to withstand a burn-up of 5 to 10 per cent of the fuel and have a heat rating of several tens of kilowatts per cubic centimetre. Liquid sodium is pumped so that it flows rapidly past them to take away the heat.

When the prototype fast reactor at Dounreay has demonstrated that it can be operated successfully, a fast reactor for the commercial generation of electricity will be designed and built. It is expected to have a capacity of at least 1,300 MW (electrical) and to be in use in the 1980s. Widespread use is then expected for this type of electricity generation. The same pattern is unfolding in other countries; France and Russia have reached a very broadly comparable stage, with the U.S., Japan, Germany and Italy following closely.

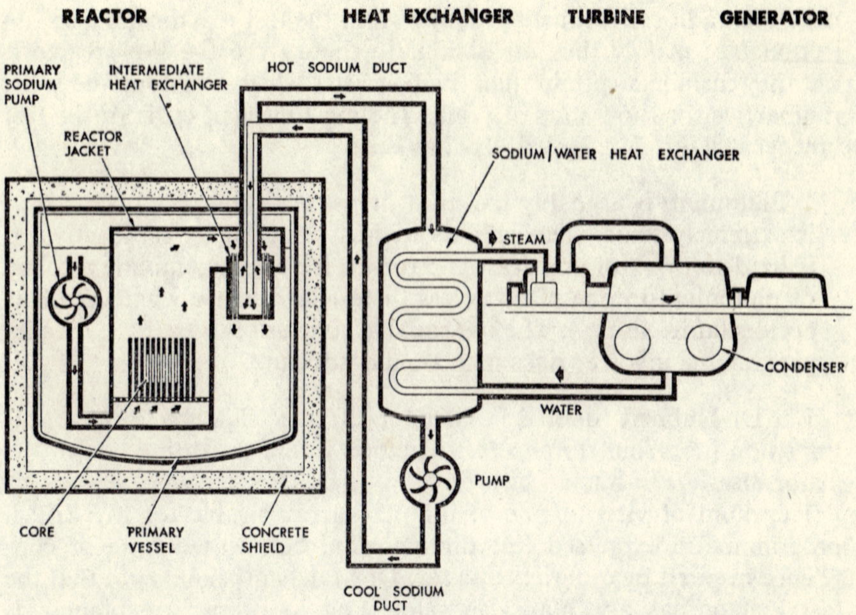

Fig. 7 Schematic of prototype fast reactor.

The studies of safety and the extraordinary measures taken to ensure safety in the event of major faults or breakdowns in operation, were important features of the early days of nuclear power. As a result the early stations, and especially the British magnox stations, have a very high reputation. It will be correspondingly more difficult for the fast reactors to reach that stage, and some critics are expressing their fears, in face of the philosophy of design of the fast reactor which puts the problem of safety foremost. Elaborate methods of shutting down the reactor (by absorbing neutrons), with two or more quite independent methods installed, are normal; emergency methods of removing heat are provided; methods of controlling and monitoring normal operation are designed to give maximum security, and containment of the reactor within a pressure vessel is an additional safeguard. A single incident would cause widespread alarm and a strong public reaction. That is understandable, if one lives with a controlled atomic bomb on one's doorstep.

Critics also point out that the fuel is 'weapons-grade' material, which is not the case with slow reactors; it is argued that a temptation is thus presented to anarchists and assassins to steal the fuel. Even without that risk, the spread of such fuel legitimately to many countries is itself a risk. The radioactive waste products from the fast reactor are also objectionable; mainly they are similar to those from the slow reactors, but they include a certain quantity of wastes which will continue to be radioactive for thousands of years. The most virulent critic of the fast breeder reactor, Jon Tinker,* points out:

> Plutonium is arguably the most dangerous substance on earth. It is extremely toxic chemically as well as being highly radioactive. It is hard to think of any adequate reason for ever assembling a tonne of plutonium in one place unless there is no chance whatever of its accidental release . . . The inescapable conclusion is that the breeder programme involves risks man should not run.

But Dr Hubbert (quoted in Chapter 13) says, 'Failure to make this transition [i.e. from slow to fast reactors] would constitute one of the major disasters in human history.'

The point of view of one of the people responsible for the British programme is expressed less dramatically, but with a tone of confidence inspired by experience so far. Dr Marsham† points out that the fast reactor has advantages in safety and, if properly exploited, is potentially the safest reactor system. He explains in detail the reasons why this is so, and also the objectives in planning, to achieve the substantial economic gains (reducing 'total system generating costs' by 10–15 per cent) that the British fast reactor programme offers. The problems to be solved, in his view, hinge on engineering, rather than on the behaviour of the fuel. Design of the commercial fast reactor of 1,300 MW electrical capacity will be finished as soon as trials on the P.F.R. permit, and it should be under construction by 1976, with a large-scale ordering programme for power stations by the mid-nineteen eighties. International collaboration on safety and on certain aspects of manufacture will enable maximum economy to be gained from the whole programme. Reliable judgments, Dr Marsham concludes, will be possible in the next few years when data from the P.F.R. become available.

Power from the light elements

The second and more important stage in the harnessing of atomic energy, namely the generation of energy from the fusion reaction, has not yet been demonstrated. It is not certain that it can be done. Nevertheless, research is being carried forward with quiet confidence and in-

* *New Scientist*, 1 March 1973.
† T. N. Marsham, 'A UK view on fast reactors', *Atom*, July 1973.

creasing competence, and I think most of the critics of nuclear physics on the one hand, and of power generation on the other, would not have serious doubts that it will be done, but maybe they would be more cautious about when it will become economically competitive.

There are two approaches which show promise, of which one seems to be far ahead in its prospects. It demands large-scale apparatus which is expensive in time and money to install, in order to demonstrate a controlled release of energy from fusion. Prudence demands that all the preliminary stages necessary to gain knowledge and experience – of which there is an elaborate but logical sequence – should be completed before embarking on the big experiment. The countries of the European community have decided quite recently to co-operate on fusion research, and it is expected that a big apparatus costing 5–10 million pounds will be started jointly by them soon. The U.S.A. and Russia have both just started work on a similar experimental project, so the stage seems set for progress towards the commercial generation of electricity from fusion in the early years of the twenty-first century.

At very high temperatures, gases exist in an ionized form called plasma. These temperatures are too high to be resisted by any materials at all, but a plasma can be created by containing it magnetically, since it is ionized, by surrounding it with magnetic fields. Confinement, as the physicists call this technique, demands a toroidal vessel, elaborately equipped with windings which create strong magnetic fields. Vessels of this type have been extensively used to investigate the physics of plasmas – previously an unknown subject, the design and properties of the windings and the magnetic fields that they produce, and methods of energizing the gas in the toroid to create – just for a moment of time – the plasma. Early difficulties in keeping the plasma stable have been overcome.

Physics of the fusion reaction

The reaction which has to be sustained to generate energy is:

$$D + T \rightarrow {}^{4}He + n + 17{\cdot}6\ MeV$$

D signifies deuterium, the isotope of hydrogen with an atomic mass of 2, and T tritium, the isotope of hydrogen with an atomic mass of 3. Helium of atomic mass 4 plus a neutron are the products of the reaction, plus a great deal of energy as expressed by the last term of the equation. An electron volt (eV) is the energy given to unit charge (i.e. the negative charge of one electron) by a potential of one volt; it is a unit of energy, not of voltage.

The plasma itself must contain a great deal of energy – about 20 keV in each of the deuterium and tritium ions – since the reaction takes place at temperatures higher than 10^8 deg.K. The energy balance input:output is thus about 40:17,600 = 1:440. Most of the energy produced – 14 MeV – is in the form of kinetic energy of the neutron. Neutrons are not charged and therefore not confined by the magnetic field. Their kinetic energy would have to be absorbed by passing them through a blanket of material, where they would collide with its molecules, so generating heat. Heat is thus the product, and the familiar paraphernalia of heat-exchange, steam raising, etc., and the equally familiar thermodynamic limitations would still be present.

Deuterium can be separated from heavy water, which occurs naturally, but tritium is not found in nature, and would have to be made. This would be done by surrounding the plasma with a blanket of lithium. Lithium bombarded with neutrons yields helium and tritium without seriously affecting the overall energy balance.

If fusion is to take place with a net gain of energy, not only must a temperature above 10^8 degrees be obtained, but the plasma must be confined (i.e. must exist) at a high enough density for a long enough time. Given the high temperature, the threshold of the reaction combining time and density is called the Lawson criterion, and is 10^{14} cm^{-3} sec. Fusion thus requires a plasma with a density of 10^{14} nuclei per cubic centimetre to be held for one second at a temperature of about 200 M deg.K. The density is equivalent to a gas (plasma) pressure of several atmospheres; to confine it, a magnetic field of about 10^4 gauss would be needed, but a greater order of magnitude, 10^5 gauss, would be a desirable practical value. For comparison, the pole windings of a 500 MW generator give a field strength in the air gap of the machine of about the lower of these two values, but the air gap is also very small.

Experimental trials up to now have achieved Lawson values of up to 10^{12} – a hundred times smaller than the threshold – but they have been done with small-sized toroids and for other purposes, as already explained. The fusion reaction in deuterium has been identified in these trials, but the plasma has cooled or been contaminated too quickly for a gain in energy.

There are various methods for creating the magnetic fields for confinement which are, without exception, complex to understand, and I will mention only that the property of superconductivity is exploited, at very low temperatures, for the windings of the magnets and that the large magnetic forces needed require currents in the range of hundreds of kiloamps in present experiments (one MA has been exceeded), all of which demands engineering of great accuracy and high quality. The most favoured magnetic system in present work is called the Tokamak; it was developed in Russia at the Kurchatov Institute, and has been improved at Culham, the British centre for fusion research. An experiment with this system at Culham which is due to begin in the second half of 1974 will investigate three fundamental matters, namely the injection of beams of high-energy neutral atoms into the plasma, in order to heat it to initiate the fusion reaction; the use of a magnetic divertor for channelling exhaust plasma within the vacuum chamber, and experimental studies of the theoretical predictions for sustained operation of the reaction, as a step beyond the pulsed operation which has been the aim of the confinement until this point.

Prospects for power

Dr Pease, Culham's director, has stated, 'We should now conduct our research on the basis that the conditions for a fusion reactor can be achieved.' In support of that conclusion, a conceptual study of the fusion power station has already been carried out for Culham, with some interesting results, summarized here but introduced by some more general points. Tokamak power station's 'fuels' would be deuterium and lithium. Present estimates of ores on land indicate sufficient lithium for 10,000 years of output at the present worldwide rate of using energy, and of deuterium, in the waters of the world, ten million times larger. There are prospects that the toroid could be given a retread, when mature, and the lithium eliminated.

The fuel would be virtually free, an advantage balanced by the cost of the magnets. In that balance lies the key to economic success. There would be no processing of fuel such as the fission power stations require, and there would be no radioactive waste products. Safety precautions would be important, but less onerous and costly, and there would be no risk of robbery of weapon-grade nuclear materials. If the reaction ran away, out of control, the risk of damage would be slight; here lies a great advantage over fission power stations, for it would seem that, given good engineering, a completely fail-safe power station could be built. The fuel (i.e. the mixture of deuterium and tritium) in the reacting zone would be only about a quarter of a gramme per gigawatt of electrical output, compared with say 3 tonnes of plutonium and 10 tonnes of uranium in the core of a fast reactor. It would be interesting in due course to assess the safety of the fusion power station in relation to conventional thermal and hydro-electric power stations; maybe it would be the safest type of all – a paradox indeed.

Figs. 8 and 9 and Table 1 explain the Culham Laboratory's conceptual

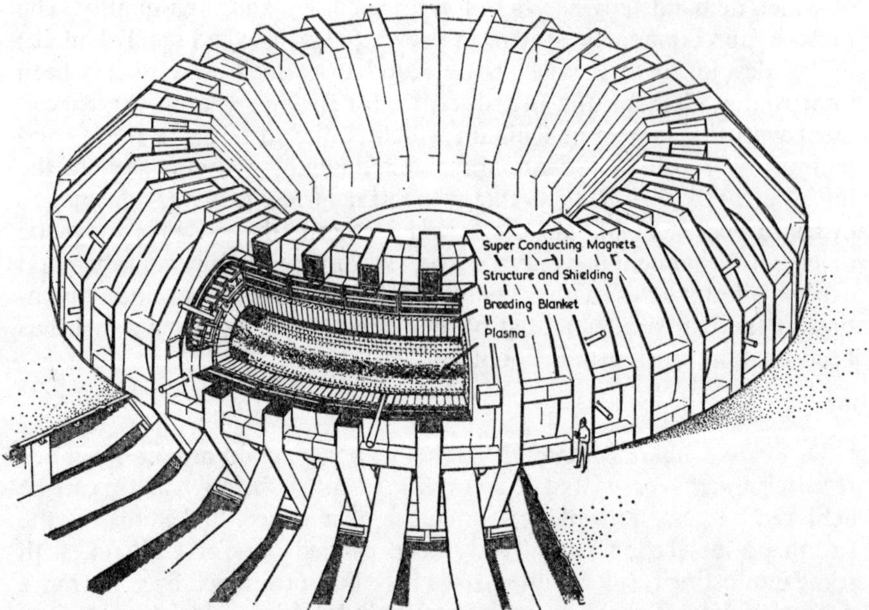

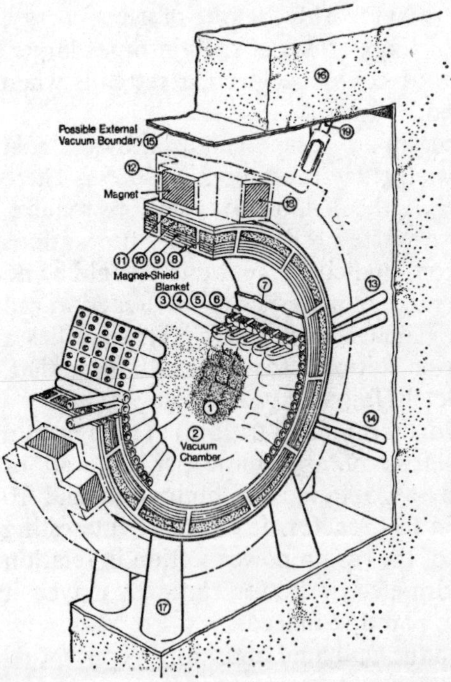

Fig. 8 Tokamak power station – how it would look.

Fig. 9 A possible arrangement of one of the 32 segments of Tokamak power toroid.

1. Plasma
2. High vacuum
3. Cellular structure containing liquid lithium (4)
5. Graphite
6. Lithium header pipe
7. Thermal insulation
8. Supporting structure (stainless steel)
9. Iron
10. Borated water
11. Lead cladding
12. Cryogenic envelope
13. Typical pipes for blanket coolant
14. Typical pipes for shield coolant
15. Lining of containing vessel
16. Containing vessel

Supports for:

17. Magnetic shield and blanket
18. Reactive forces of magnets (compressive)
19. Weight of magnets (tensile)

(Source: U.K.A.E.A.)

study* for a power station with an electrical output of 2,500 MW. Fig. 8 shows the toroid, and Fig. 9 one of its rings; it is envisaged on the unit principle, whereby a ring could be taken out and replaced if maintenance were needed.

The speed of the neutrons determines the thickness needed for the blanket of lithium, and this dimension sets the scale of the toroid. The windings of the magnets are outside the toroid, and would be super conductors cooled by helium to 4 deg. K, and carried by a structure within the cryogenic insulation which could resist the large forces exerted magnetically between the segments. Refuelling, probably combined with injecting starting energy, removing impurities from the plasma, magnetic divertors, start-up windings for plasma current, these are functions not yet worked out.

A reduction in price of ten to thirty times the present prices would give reasonably competitive costs for the magnets. That could be possible; the physical properties demanded of these materials are less certain of achievement than a reasonable price.

There is a square-cube relationship between cost and output. The size chosen for study takes account of the well-known criterion of 10 per cent of the night-time summer base load on the electricity grid, which is about the maximum size that a power utility will install in one unit. The advantage of a much larger size has posed the question 'What is the biggest source of heat the electricity system can accommodate?' The answer leads to consideration of, say, a circuit of hot liquid lithium from the great toroid, serving by heat exchange some arrangement of boilers and hence turbine-generators. The penalty in capital cost for exploiting the advantage of size would be a group of standby heaters, presumably oil-fired, which would maintain the lithium in circulation if the toroid came off load by breakdown of some kind. This is the genesis of the United Kingdom Atomic Energy Authority's concept of generating power at a few large centres by the end of the century. The only thing left out of the concept appears to be a proposal to apply the waste heat from the turbines to useful purposes, but that is an old story that Tokamak power station has inherited, and the increase in scale appears to make such an application less likely than ever.

The next step along the path will be taken by the large-scale experimental toroids. They will investigate conditions at high temperatures; when consistent and steady 'flame' conditions are possible there will be a further stage in the evolution of plasma physics. With that knowledge and experience available, a prototype heat generator can be designed

* See also R. Hancox, *Fusion Reactor Research Studies in the United Kingdom.* To be published.

and built. It would probably be in the range of size of 100 to 500 MW of thermal output (thermal output would be at least three times greater than electrical output) and there would be advantage in speed and economy in keeping the prototype small, but disadvantage in that it would then be farther removed from the production job, which for economy would probably have to be very big.

Now a complete change of subject. I mentioned a second approach. It envisages the collapse or implosion of mixtures of deuterium and tritium to very high densities by irradiation with lasers of high power. Energy has to be supplied at very high rates, and densities greater than those at the centre of the sun would be necessary; at that point the fusion reaction would be initiated. Lasers of 160 J are available; 10 kJ would be needed for experiments, and 1 MJ for a reactor. This method is being pursued mainly in the U.S.A. and Russia. It has great basic scientific significance, and is not confined to the reaction between deuterium and tritium. Teller has said that this approach could form a basis for the propulsion of space vehicles. As for direct generation of electricity (a reaction between deuterium and helium 3 gives a charged particle) prospects are hazy.

TABLE 1: TOKAMAK POWER STATION: A CONCEPTUAL DESIGN

electrical output	2,500 MW
plasma vessel:	
major diameter	25 m
internal diameter	5 m
number of segments	32
mean rate of heat removal at wall	4·6 MW/m^2
circulation of liquid lithium	4 tonnes/sec
superconducting magnet:	
mean field	95 kG
plasma current	10 MA

A steady-state fusion reactor, the plasma confined by a Tokamak toroidal-pinch magnetic system. Deuterium–tritium fuel is consumed at a rate of 50 grammes/hour. Heat is generated in a blanket of liquefied lithium which also serves to breed tritium.

The diagram of one segment (Fig. 9) shows that the blanket is cellular, and is backed by a radiation shield of iron and borated water. Typical thicknesses of blanket and nuclear shield are 65 and 75 cm respectively. The plasma exists in a toroid about 1·5 m in radius, and the first wall – separated from the plasma by a high vacuum – is at a radius of 2·0 m from the centre of the toroid.

6 An Arcadian idyll

The pure dew on the grass, and the crystal water of the river, rippled by a virgin zephyr's breath: are they luxuries that the affluent world of today cannot afford? Let us examine the categories of muck and poison thrown off by energy usage, but ignore the personal exhausts of the dirty, wasteful, but above all numerous, population of this planet, and also wastes from industry not directly caused by the usage of energy.

Pollution of the atmosphere by burning fuels, domestically, industrially and in the internal combustion engines of motor vehicles makes an obvious starting point. Carbon (coal) burns to carbon dioxide (CO_2), a gas that is present in the atmosphere in small amounts. Hydrocarbons (oil and natural gas) burn to CO_2 and water. Incomplete combustion gives the monoxide of carbon (CO) in place of some of the dioxide, which is highly poisonous. Where high thermal efficiency is sought incomplete combustion is an indication of bad control, and with fuels costly and competition severe, is not widespread in modern industry. It is universal, however, in the internal combustion engine. Volatile fuels (hydrocarbons) may also escape into the air without being burned at all.

Sulphur is generally present in coal and oil, and in some natural gas. It burns to the dioxide or trioxide (SO_2 or SO_3), both of which are acidic and corrosive when they combine with water in the atmosphere. They are unpleasant and injurious to health, and deaths caused by polluted air have been attributed mainly to these gases. Oxides of nitrogen also occur in the exhausts from combustion, mainly by oxidation of the air, not the fuel. They too are acidic; small quantities occur naturally. Acidic products of combustion blight vegetation and cause rapid weathering and decay of buildings, a process in which soot is their ally.

Ash and soot are substantial exhaust products. There are a few others, such as finely divided lead, which is exhausted by motor vehicles. The typical industrial pollution consists of a smoky, sulphur-laden atmosphere, maybe given a special character by some local industry. The motor vehicle may have different effects, and it is convenient to consider the two separately.

Oil is a less offensive fuel than coal in terms of atmospheric pollution. This comparison cannot be made for the internal combustion engines of motor vehicles, which are the source of today's most urgent problems, but in this case one can blame the engine rather than the fuel. Even as an industrial fuel, however, if one takes account of the pollution at the refinery as well as the point of combustion, oil loses some of its advantage. It is not easy to generalize about the smoke, sulphur and smells from refineries. Waste products are burned off; the smells are nasty but appear to be harmless; control of pollution should be effective. Yet some of the refineries I have seen have been swathed in satanic murk and flames, and it has been very relaxing to leave them behind on the horizon.

Bituminous coals contain sulphur mostly from 0·5 to 5 per cent (average 1·7 per cent in Britain) by weight. Most crude oils contain less than 1 per cent by weight, but 5 per cent is not unknown. There is sulphur in some natural gas, generally as hydrogen sulphide which can be removed, but like the sulphur in coal and oil it may then be burned to the oxide and discharged to the atmosphere. It is impracticable to remove SO_2 from flue gases.

TABLE 2: U.S. STANDARDS FOR ATMOSPHERIC POLLUTION

	SO_2	Particulate matter	CO
Hazardous concentration ($\mu g/m^3$)	2620*	1000	86,300
for exposure for time of: (hours)	24	24	2
Primary standard ($\mu g/m^3$)	80 mean	75 mean	40,000 1 hr max
Irritation of nose and throat ($\mu g/m^3$)	15,000–30,000	—	—

* Equivalent to about one part per million. Detected by taste, 0·3–1·0 ppm; by smell, 3 ppm.

In Britain about 5 or 6 million tonnes of SO_2 is discharged each year mainly from industrial sources.

It is difficult to make a simple generalized statement about the effects of SO_2 on health, except that they are bad, or maybe very bad. People with respiratory troubles are more vulnerable than others. Table 2 gives in outline only, some leading figures for standards of cleanliness (primary standards) or of danger, as laid down by the U.S. Environmental Protection Agency, in this case living up to its name literally, and protecting one from the environment. The hazardous concentration is defined as 'might cause significant harm to human health' but other U.S. standards give ten times that concentration as permitted for eight hours' exposure in working areas, which seems much too high in terms of the Agency's definition and figures.

Atmospheric wastes from power stations

The usual practice is to pulverize coal for power stations, and then to inject it into the boiler as a powder. In principle one might expect a lot of soot from this practice, but from the point of view of efficiency soot is worse than carbon monoxide since it is unburned fuel, and in fact combustion is carefully controlled, so finely divided ash, known as pulverized-fuel ash (pfa or fly ash in the U.S.A.), is the only non-gaseous product of any consequence. Dark smoke may be seen at start-up or if the boilers are banked. A white or off-white plume consists of steam plus residual ash. Steam is formed by condensation as the hot gases cool, or maybe occasionally the direct rejection of steam by the power station.

Nearly all the ash is trapped by electrostatic precipitators. Only a small percentage – about 1 per cent – gets through these devices. Although small as a percentage of the total, the quantity is large. A power station burning 2,000 tonnes of coal in one hour will make say 130 tonnes of pfa in that hour (say 8 per cent ash, of which 80 per cent is pfa) and about 1½ tonnes of it will escape from the chimney. The quantity of CO_2 emitted will be about 6,000 tonnes, but it is unimportant and indeed unnoticed. Sulphur to the extent of, say, 2½ per cent by weight gives 50 tonnes of sulphur, equivalent to 100 tonnes of sulphur dioxide, debouched in the hour. The SO_2 will have to be dispersed uniformly in a volume of 40 cubic kilometres if its concentration is to be reduced to the U.S. hazardous standard, and 1,250 cubic kilometres to dilute to the primary standard.

There is only one unsubtle method of diluting the SO_2 and ash by the enormous numbers of times that are essential. That is to build a tall chimney, 200 or 300 metres in height (Plate 4). The gases emerge at a velocity of about 6–15 m/sec which is quickly reduced by cooling from the air all round, and then mixing takes place. But under some conditions of wind and atmosphere, the plume from a power-station chimney can be depressed, in a relatively concentrated form, and find its way to ground level before dispersion has been effective. Generally, the higher the chimney, the farther afield it spreads its unpleasant dejecta of ash and sulphurous fumes (Plate 4), but the generality is neither universal nor precise, although of course there are design criteria which are applied to give results in terms of probabilities which seem to give reasonable local protection around the great power stations that have been built in recent years.

That at least was my view; maybe I'm out of date. Here are the conclusions of three specialists:*

> A high chimney is a cheap, reliable and indeed indispensable means of reducing pollution by gases. The criticism of chimneys is largely

* A. J. Clarke, D. H. Lucas and F. F. Ross, 'Tall stacks – how effective are they?' Central Electricity Generating Board. Second International Clean Air Conference, Washington, December 1970.

misguided. The critics tend to consider the atmosphere as a whole and do not consider in sufficient detail the small part of the atmosphere in which we live and breathe. It would be a tragedy for the cause of clean air if the campaign for low-sulphur fuels and sulphur removal led to an underestimation of the potentiality of chimneys.

No sulphur dioxide removal process – even if 100 per cent efficient – can make it possible to dispense with a chimney. A chimney can in most cases do all that a removal process could do at a fraction of the cost to the community. An efficient removal (or prevention) process is required for smoke and dust. There may also be a case for avoiding unnecessary nitrogen oxide formation. But a removal process for SO_2 would only make sense where the adverse circumstances of terrain or a nearby airport, or the social desirability of making the best use of high-sulphur fuel, make the highest practicable chimney inadequate. Even then it would only be needed for comparatively infrequent periods, need not be more than partial unless the process cools the gases, and should still not be considered until after potentially cheaper means of reducing pollution during these periods have been rejected as impracticable.

The authors believe that the toxicity of sulphur dioxide has been exaggerated. Sulphur is present in coal partly because it is essential to plant life. If the ground-level concentrations of sulphur dioxide are sufficiently low the gas not only becomes harmless, it may be beneficial.

The authors also believe that much of the aesthetic objection to tall stacks is rooted in the observation that far too many chimneys have in the past, and still do, emit unnecessary quantities of smoke and visible particulate matter. Some spokesmen for clean air have been over-critical of tall stack dispersion of the invisible constituents and have encouraged the public to believe that this control measure is neither adequate nor acceptable. This is contrary to the public interest. It has caused, and continues to cause, the expenditure of engineering man-hours and huge research funds on a problem for which an answer exists, thereby diverting them from the real problems of air pollution; these are the emission of sulphur dioxide in cities and, in some areas, the emission of unburnt hydrocarbons from vehicles.

Pollution is not merely an industrial problem, it is also a neighbourhood problem. Citizens who point critically to the power station burning three per cent sulphur oil with an adequate chimney, and ignore the 0·3 per cent of sulphur in the heating oil used by themselves and their neighbours, do not understand the facts. If citizens believe that SO_2 is harmful to them and consider that its concentration in the air they breathe should be reduced, then the remedy is the most stringent restriction on the sulphur content of the fuel they themselves use in their daily lives and the provision of adequate chimneys for large industrial sources.

Motor vehicle exhausts

The most serious cause of atmospheric pollution in most cities today is the motor vehicle. It exhausts CO_2, water and nitrogen from the fully-burned part of the air and fuel which it consumes, but also, from incomplete combustion, carbon monoxide, carbon (soot), hydrogen, oxygen, unburned and partly oxidized hydrocarbons which are constituents of the fuel or the lubricating oil, and oxides of sulphur and nitrogen. Additives which give special properties to the fuel result in further exhaust products, principally lead.

Large amounts of CO from incomplete combustion have been inevitable up to now in the exhausts from motor vehicles. CO is without smell and cannot be detected by one's senses, but is highly poisonous because it interferes with the intake of oxygen by the lungs. That is why running a car in the garage with the doors closed is an efficient way of committing suicide. It is surprising that pedestrians in highly-trafficked city streets are not regularly overcome by CO poisoning; clearly the atmosphere has great powers to disperse and dilute. Even so, one reads that in Tokyo, policemen go on duty for one hour only where traffic is dense because otherwise they would get poisoned. Recently a case occurred in Washington, D.C., where a policeman on traffic duty collapsed; in due course his relief also collapsed. Carbon monoxide poisoning was the stated cause, and comment at the time included the assertion that children now growing up in the city had little prospect of living a full span of life without some sort of trouble in the lungs.

Inside the car, if there is some leak from the exhaust, or even if the air is taken in for heating or conditioning and the car is caught in a heavy jam of traffic, the occupants may be more vulnerable than the pedestrians, and several accidents have been blamed on partial loss of faculties by the driver caused by carbon monoxide poisoning.

The quantities of the various exhaust products, and specifically CO, depend to some extent on how the engine is running. Taking a city street 100 metres long with four traffic lanes, jammed solid with traffic with all engines idling, around half a cubic metre of CO would be generated in one minute, which would need to dissipate over the full width of the street (20 m) and to a height of about 10 m before its average concentration reduced to the one-hour-exposure limit of the U.S. primary standard.

In a recent lecture on air pollution* Sir Eric Ashby, chairman of the Royal Commission on Environmental Pollution, pointed out that more CO was put into the atmosphere – 230 million tonnes a year, worldwide, 80–90 per cent of it from motor vehicles – than any other gas except CO_2. Brief periods of concentration in city streets produced 15 per cent saturation of the blood (at which symptoms occur) but the gas diffused away quickly. Concentration of the gas in the atmosphere had not changed because micro-organisms in the soil were scavengers of CO, and converted it into methane and CO_2.

Apart from this directly poisonous habit, the motor vehicle has created an entirely new pollution phenomenon, photochemical smog. Just as London is the ancient home of the traditional peasouper – a sort of genuine beefeater smog – so Los Angeles can claim to be the home of the up-to-the-minute trendy version, photochemical smog. People catch on so fast these days that it can be accurately recorded that latter-day Sherlocks in dear old London have also detected the new variety – a little only yet, but it has arrived.

Photochemical smog, as its name implies, is formed by the action of sunlight on hydrocarbons and oxides of nitrogen. Both these classes of substances are exhausted by motor vehicles and when they are trapped by an atmospheric inversion and concentrations build up, when the sun shines, smog is formed. It smells and causes irritation of the eyes.

Tetra-ethyl lead is an anti-pinking additive in petrol. Its diffusion from vehicle exhausts is given as the only reason why lead can be detected in the atmosphere in remote, uninhabited places. Lead can poison people, so these measurements have caused alarm, but there seem to be more immediate possibilities of lead poisoning for most people – for example lead water pipes in soft water areas – and other measurements have shown lead in the hair of old corpses that has been taken as evidence that they had more lead in them than we have. How comforting!

Fine soot is also prolific in vehicle exhausts. Diesel engines tend to be worse than petrol, although only if the fuel injectors are not adjusted, which from common observation seems to be quite often. But the relatively invisible soot from petrol engines, apparently from the lubricating oil, has been said to be more harmful.

Motor vehicles have other attributes which present more serious problems than pollution, but few people are prepared to do without the speed and convenience of travel which motor vehicles offer, and their numbers continue to grow.

Noise is now classified as a pollutant. Electric cars could be on the verge of invading us because they are silent and exhaust-free. There is

* Ernest Balson Lecture, Institution of Public Health Engineers, January 1973.

a trade-off here, just as with oil and smokeless fuels; electric cars would increase the demand on power stations and transfer exhaust from the city streets to their chimneys.

Pollution by aircraft

The aeronautical jet engine is a dirty beast. I remember at Farnborough in the fifties, the commentator enthusing over British jet engines because of their clean exhausts, and criticizing American jets, saying that if you followed them with a bucket you would keep on getting it filled with paraffin. Alas, with increasing performance and thrust, they all began to leave a trail of black smoke.

National Airport at Washington, D.C., is a good example of the destructive ambience which these graceful and speedy machines create. Aircraft approach or take off along the line of the Potomac, low over the city, which they shower with soot and paraffin every few minutes. Because noise has to be reduced to a deafening diminuendo, more fuel is used and a plume of black smoke is inevitable from each aircraft. The Lincoln Memorial lies in the shower bath, and its great marble bulk has recently been treated at enormous cost – and without much hope of success if I may extrapolate rather incautiously from the stated experience of the British Building Research Establishment in similar matters – to stop it crumbling away.

These conditions, however, annoy only in the neighbourhood of airports. The newest aircraft, like Tristar, are less noisy and free of soot, financial soot excepted. The predictions are that aircraft operated by the main airlines will be quiet by the 1980s. Jet aircraft fly very high and make dirt and water vapour at high altitude. A doubling of high cloud in ten years has been reported from Denver, Colorado, presumably because of these emissions from high flying aircraft. Concorde flies even higher, at 15 to 20 kilometres above the ground, and it has been suggested that its exhaust might destroy ozone in the upper atmosphere, thus doubling ultra-violet radiation at ground level, with various disastrous consequences. This theory has been hammered somewhat vigorously, but persists.

The Clean Air Act

Ships as a source of atmospheric pollution can be ignored, because of the vast spaces in which they travel. Railway locomotives perhaps too, because they are not very numerous, although it can be very unpleasant near a big diesel loco in a covered railway station. I was rather anxious to quote in this chapter an old saying about the steam locomotive, that

* R. S. Scorer, 'Could Concorde fry the earth?', *Engineering*, August 1971.

it was the most efficient machine for polluting the atmosphere that had ever been invented. Before the words could reach the paper, I was reproached. Steam locos only emitted smoke when they were being badly stoked, I was informed, and otherwise were clean. The greatest maker of smoke in the days of coal and steam was the Lancashire boiler, which had no long tubes to arrest soot, and happily spouted great clouds of black smoke up each and every industrial chimney in the land.

Maybe those days have passed, but pollution from coal burned inefficiently is not entirely a thing of the past. The third appreciable source of atmospheric pollution remains the domestic consumption of fuel; rather inefficient combustion from a multitude of small, single-house burners can be a major source of sulphur and smoke. Large central heating and district heating installations tend to take on an industrial character, and the growth of such installations combined with, in Britain, less dependence on coal fires, has alleviated very considerably the pollution from this source. There have, however, been other more direct factors at work.

The Clean Air Act of 1956 has been the legal impetus promoting these direct factors. It set out to establish zones of clean air – 'smokeless zones' – in which it is an offence to emit black smoke. The Act was aimed primarily at the pollution caused by burning coal. At that time it was estimated that half the atmospheric pollution in Britain was caused by domestic chimneys serving open coal fires. Special fuels were introduced or encouraged which were smokeless or nearly so, as were smokeless methods of lighting domestic fires. Both were mandatory in smokeless zones, where smoking industrial chimneys and vehicles emitting black smoke also became illegal. Smokeless zones were established gradually, and are still being extended.

The Act was a pioneer in the modern world – the first of its kind anywhere. But it had its precursors in medieval London, which was cursed with 'insufferable smoke'. Towards the end of Elizabeth I's reign, the burning of coal in London was prohibited while Parliament was sitting – maybe soot would have got into the legislation. The Act of 1956 came about because of increasing concern generally, but above all because of the infamous smog of 1952. The whole of Greater London was blanketed with an acrid, dense fog, trapped in the stillness of a temperature inversion of the atmosphere. It lasted five days, and 4,000 deaths, mainly of old people or people with respiratory trouble, were attributed to it. At the time I lived at Barnet, which in its highest parts is about 125 metres above sea level. Walking on Hadley Green, the highest part, one looked up and saw the sunshine just reaching the tree tops, but looking horizontally, the fog was dense. Box Hill in Surrey is higher, and protruded above the smog, a fact which *The Times* displayed photographically at the time. One fact which emerged was that power stations did not contribute very much to the smog. It was proved by aerial photography that the plumes from their chimneys penetrated above the inversion, as can be seen in the picture (Plate 8).

Ten years later, in 1962, another smog occurred, but less than a tenth of the number of deaths was attributed to it because of increased public awareness of the dangers (people stayed indoors and didn't burn coal or run motor vehicles) but also because of the Clean Air Act. The act has been applied slowly, and has powerful critics in spite of the achievement shown in the graph (Fig. 10). An objective assessment of its influence is this:

> The air over London is substantially cleaner today than it was fifteen years ago. There has been an 80 per cent reduction in smoke emission, a 40 per cent reduction in sulphur dioxide, and a consequent near doubling in the average hours of winter sunshine. It is estimated that this dramatic improvement – largely the result of the enlightened Clean Air Act of 1956 – has cost Londoners only about 35 cents per annum. What it has saved them in discomfort and illness is beyond calculation . . .*

Recent comment on the Act includes a piece by Bradford Imrie† which comments on the standard of accuracy of the methods of measuring air pollution specifically in Leeds, which city is described in a gem of journalese as 'an air pollution black spot'. The graph (Fig. 11) shows the official figures for Leeds; it is worth comparing these quantities with the U.S. standards quoted on page 54, and drawing attention to Mr Imrie's comments on them which are very pertinent. He says that results have been published which show that exposure to the U.S. primary standard for SO_2 for a year continuously could give respiratory trouble to some people, and brief exposure to higher concentrations could cause irreparable damage to the lungs. The graph, he suggests, is not representative of the whole of Leeds, and conceals monthly and daily variations. Highest daily mean smoke and SO_2 concentrations of 643 and 921 $\mu g/m^3$ respectively were recorded at one of the nine sampling sites. His conclusions: care and diligence is necessary in sampling, and urgent action is necessary to remedy the pollution; the Act does not control SO_2 and its reduction occurs because of the control of smoke; legal limits for SO_2 are essential.

* Robert S. McNamara, President, World Bank Group. Address to United Nations Conference on the Human Environment, Stockholm, 1972.

† *New Scientist*, 4 January 1973.

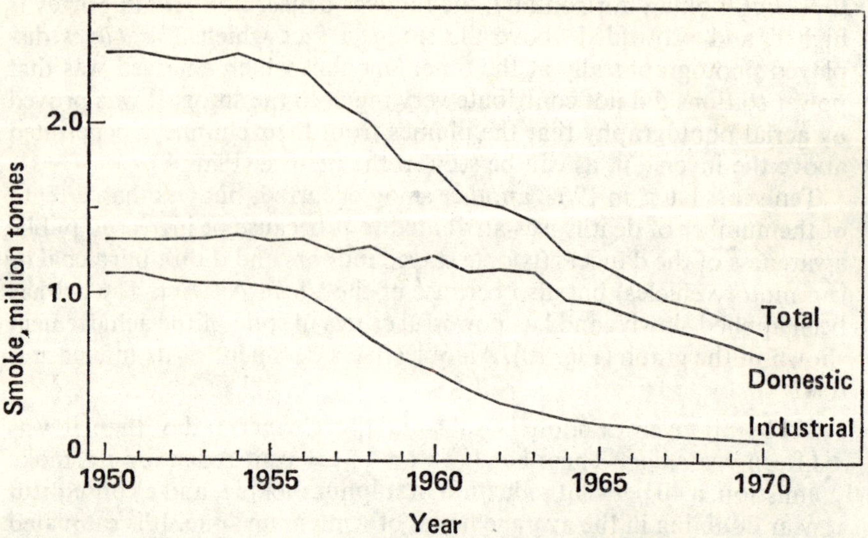

Fig. 10 Smoke over the United Kingdom reduced by 65 per cent between 1954 and 1970, despite a 9 per cent cent increase in population and a 29 per cent increase in gross energy consumption.
(Source: Warren Spring Laboratory.)

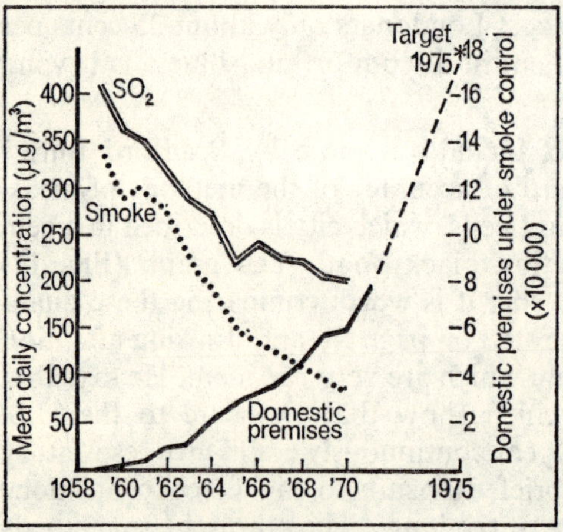

Fig. 11 Official figures showing reduction of concentration of smoke and SO_2 in Leeds, compared with the total number of dwellings in the city's smokeless zones. The scale for numbers of dwellings under smoke control is on the right.

Research has recently been started on the behaviour and distribution of SO_2 in the atmosphere. Early indications are that SO_2 is unnoticed in natural conditions because it oxidizes quickly and is deposited. It is likely that Scandinavia's recent complaints about pollution from British power stations are unfounded, and that SO_2 is soon deposited out of the air as sulphate.

Everything has its price. I recently saw a description of a plant which makes smokeless fuel. It was emitting yellow smoke which was killing the trees round it. Gas burners were installed in the flues to burn the smoke. So one is, in the first case, exchanging only the site, and hence the concentration of pollution. This in itself is worthwhile because pollution consists of concentrations, not absolute amounts. In the second case, one is burning off smoke to waste, and using extra high-quality fuel to do it. It would be interesting to see an energy balance sheet for this operation and assess just what is the price, in terms of wasted heat, of destroying the smoke.

The 1970 Clean Air Act of the U.S.A.

I described the British Clean Air Act of 1956 as a pioneer of modern times. Beside the U.S. Act of 1970 of the same name, it looks like a Newcomen engine standing beside a gas turbine. The U.S. Act is a phenomenon of interesting novelty, and is more of an industry than a legal dictat. It would seem that the U.S. is going to get clean air. Here is a brief and simplified résumé of how it is being done, glossing over factors like the relationships between federal and state authorities which are necessary but domestic in character, and noting in passing that any U.S. citizen can take action to enforce the Act.

The purposes of the first part of the Act are: 'to protect and enhance the quality of the nation's air resources . . . to initiate a national research and development program . . . to provide assistance with air pollution prevention and control programs and . . . to encourage regional air pollution control programs'.

As a first step then, research, investigation, and training are actively promoted, to form the foundation of educated appreciation, criticism and day-to-day control which must complement legal enforcement if success is to be consistently achieved. Research is aimed specifically at developing cheap methods of controlling the products of combustion and the evaporation of unburned liquid fuels, removing offending substances from fuels before they are burned, improving the efficiency of combustion, and developing new fuels which will pollute less.

National standards of quality of the air – primary (see page 54) and secondary – are specified and each state has to submit a plan to the federal government showing regionally how it is going to achieve, maintain, and enforce the chosen standards. The Environmental Protection Agency's (E.P.A.'s) administrator has to publish and revise lists of air pollutants and criteria for quality relating to them and, initially in 1970, proposed regulations to achieve primary or secondary standards, which also can be revised. Proceedings for consultation, adjusting timing, and enforcing are defined.

'Stationary sources' are the first category of polluters considered in the Act. Enforcement goes as far as a maximum fine of $50,000 per day or imprisonment for up to two years after the first conviction. There are powers of inspection, monitoring and entry. In certain cases, a conference may be called for the purpose of abating pollution.

The Act has attracted most interest for its severity in relation to motor vehicles, and this is the topic on which most of the comment has centred. This second category of polluters is called 'moving sources' and the second part of the Act which deals with them is referred to also as the National Emission Standards Act. Its principle has been to push the manufacturers hard to improve successively, as each new model is put into production, and the legal standards relate to the year in which the car was made.

Controls actually started with the 1968 models, but stricter standards were introduced for 1970 models. The Act states that for 1975 models a 90 per cent reduction in the emissions of CO and hydrocarbons compared with the 1970 models must be achieved. Further, that 1976 models will emit 90 per cent less oxides of nitrogen than 1971 models. The manufacturers asked for a year's delay in the introduction of these standards which has been allowed.

The new, strict standards are also rather more simple to tabulate. They specify maximum emission per vehicle mile, irrespective of the size of the vehicle's engine. Light-duty vehicles (that is, passenger cars that carry twelve people or less, or vans weighing 6,000 lb or less) are covered whether petrol- or diesel-engined. They must, beginning with the 1975 models, emit not more than the amounts specified in the table below.

TABLE 3: MAXIMUM PERMITTED EMISSIONS FROM LIGHT DUTY VEHICLES (U.S.A.)

		grammes per vehicle mile	approximate equivalent
CO		3·4	0·15%
hydrocarbons		0·41	33 parts per million
oxides of nitrogen:	1975 models	3·1	800 ppm
	1976 models	0·40	108 ppm

Compared with the street jammed full of cars used as an example on page 57 in which we took cars not subject even to the U.S. standards for 1968 models, the CO they generated in one minute would, at 1975-model U.S. standards, be generated in about 30 minutes.

The regulations for heavy-duty vehicles remain more complex. They were first applied to the 1970 models, and in September 1972 tighter standards applicable to 1974 models were announced. They are given in the table opposite. Clearly, slower progress is being made with heavy vehicles, but these standards lower CO by 30 per cent, introduce control of oxides of nitrogen, and reduce smoke by about half compared with the preceding ones.

Having specified, the problems start to heap up. First, how is it to be done? Then, how is the standard to be checked, maintained and enforced for vehicles in use? There is still a great deal of movement in the first of these topics, and clearly manufacturers may not reveal their methods, which could include many possible variations in design or

TABLE 4: MAXIMUM EMISSIONS FROM HEAVY-DUTY VEHICLES (U.S.A.)

1974 models	grammes per brake horse-power/hour	equivalent to
CO	40	1%
hydrocarbons and oxides of nitrogen, combined measure	16	1,800 parts per million
	% opacity of plume	
smoke emission: acceleration	20	
uphill	15	
maximum any time	50	

operation of their engines or in the constituents of the fuel. There are, however, for the light vehicles two principal methods which have been applied. In the commonest method, an afterburning device is used with a catalyst, thus oxidizing CO and hydrocarbons. Further modification – the dual catalyst method – is necessary to control oxides of nitrogen. The exhaust gases first pass through a reducing stage where a catalyst reduces the oxides of nitrogen to nitrogen or nitrous oxide, then air is introduced and the second catalyst is put to work to oxidize the CO and hydrocarbons. Lead-free petrol is essential because lead poisons the catalysts.

The second method depends on refining the design of the engine so that it burns the fuel completely, without any catalyst at all. It is exemplified by the Honda Motor Company's compound engine in which vortex flow of gases gives, eventually, complete combustion. A rich mixture of air and fuel is burned in a 'pre-chamber' and the burning gases expand into the main chamber of the cylinder, where a lean mixture – too lean to be ignited in a conventional engine – is completely burned. Oxides of nitrogen are controlled because a high temperature is needed to form them, which can occur only in the pre-chamber where the quantity of nitrogen is small.

In its 'state of the art' summary at the end of 1972 (remember it has since all been delayed by a year), the E.P.A. reported three non-catalytic models, all foreign, up to 1975 standards and 'probably able to certify' for 1976, i.e. 90 per cent reduction of oxides of nitrogen. General Motors, Chrysler and Ford, representing 85 per cent of the American market, would be 'probably able to certify' for 1975, but were 'judged questionable' for 1976. The conclusion was:

> If E.P.A. does not adopt averaging of the emissions from vehicles with a given family of engines for its testing, then only one manufacturer (Honda) would be predicted to even have a chance of complying with all the emission requirements for model year 1975. If E.P.A. adopts the averaging of emissions then at least 85 per cent of the current American market will probably comply. . . . The major problem with the 1976 systems is the absence of an acceptable NO_x catalyst. The three non-catalytic systems have potential to meet the 1976 standards. The catalytic emission control system that has so far achieved the 1976 emission levels for the greatest number of miles is not a typical dual catalyst system and is being developed and demonstrated by a supplier to the automobile industry. The on-the-road emission from 1975 vehicles may be somewhat higher than the standards against which prototype vehicles are certified, because vehicles may not be maintained and operated carefully enough to meet the standards. This issue will become far more important for 1976 systems, because emissions of hydrocarbons and CO from a 1976 system may even exceed those from uncontrolled vehicles if catalytic control fails in service.

The E.P.A. steers clear of the 'legal and policy implications' of averaging the results of tests, but points out that the regulations extend to a useful life of 50,000 miles, but in fact 85,000 miles is the average. The habits of the average motorist, and the influence of inspection in changing them, will influence the eventual success of the Act, in addition to the technical progress which will have to be made in the remaining time before the new standards are introduced.

The complexity and effort which success demands will by now be abundantly clear. The regulations for cars are encyclopaedic. But there seems to be determination to succeed. Bypass devices which can shut off the emission control system have been prohibited; tampering with lead-free gasoline carries draconian penalties: the first recall order was made on 15 February 1973, when 2,290 General Motors' cars – the entire production run of these particular models – were ordered back to Detroit for modifications. In the same month a dealer in Florida was charged with 'rendering inoperative' the pollution control system on a 1972 car; the maximum penalty is $10,000 per violation.

But the E.P.A. itself seems to have neatly capped the dilemma of the 1976 models. In June 1973 it stated that 'nitrogen dioxide (NO_2) does not represent a widespread pollution problem' and that the 'mandatory

90% reduction in auto emissions of nitrogen oxides by 1976 is not needed'. In 1974 it will devise a new 'reference method' for measuring oxides of nitrogen.

Carbon dioxide

The most abundant product from burning fuels is carbon dioxide. Enormous quantities of this gas are thus being generated. The quantity of it in the earth's atmosphere is increasing, and will grow quite rapidly, according to the best forecasts, if the burning of fuels continues to accelerate at the same rate as in recent years. This increase seems to have been overlooked by the more vocal and hypersensitive environmentalists, which is surprising.

The atmosphere consists roughly of four-fifths nitrogen and one-fifth oxygen, with about one-thirtieth of one per cent of CO_2 and nearly one per cent of other gases, mainly the rare gases, which are inert. Carbon dioxide has no direct effect on health, makes a contribution to some geophysical processes of weathering, but has no major effect at all except perhaps – or noting the degree of precision of the argument which follows, better to say probably – on climate.

Energy is radiated from the earth, so causing a lowering of the temperature of its surface, in the form of infra-red light. Some of this radiation is absorbed in the atmosphere by clouds and water vapour, and by carbon dioxide. So with more carbon dioxide there would be a tendency for climate to grow warmer. It does not seem that any reasonably accurate prediction of whether or not such changes will be significant has been made, or can be made with present knowledge. Nevertheless one scientist has described this phenomenon as 'man's greatest geophysical experiment'.

Analysis separates the problem into three components. First, for a given increase in CO_2 in the atmosphere, what the effect on climate would be. Second, the dynamic balance between any such increase and the vast reservoirs of CO_2 in the oceans and in living things and soils. Third, the likely range of 'production' of CO_2 in the coming decades.

Energy radiated off the earth depends on the earth's temperature, snow cover, and the opacity of the atmosphere to the radiation. About a third of the energy received by the earth as radiation from the sun is directly re-radiated off again. Opacity of the atmosphere can reduce both the incoming and outgoing radiation. Cloudiness, water vapour and CO_2 are the relevant variables.

If only the CO_2 varies, then calculations suggest that a doubling of CO_2 will give a rise in mean global temperature in the range 1 to 1·5 deg.C. This result is useful as a guide, but in practice there would be a dynamic interplay between the various factors. Higher temperature could also increase water vapour and cloudiness, hence magnifying the increase caused by more CO_2 alone, but this is not certain, and just what the new dynamic balance would be is too difficult to forecast. Changes observed in climate in the past are not significant when related to past increases in CO_2.

Nearly all the earth's CO_2 is dissolved in the oceans. Living plants and the organic processes of soils contain more than twice the quantity in the atmosphere. According to Broecker *et al.**, the amounts, measured as carbon, are as in the Table below.

TABLE 5: TERRESTRIAL CO_2

	Gram-mols of carbon $\times 10^{16}$	per cent	
Atmosphere	5	1·5	
Biosphere	14	4·0	
soils			3·2
plants			0·8
Ocean	325	94·5	
warm surface			1·5
thermocline			23·3
deep ocean			69·7

The living processes 'fix' atmospheric CO_2 by photosynthesis but it seems unlikely that increased CO_2 in the atmosphere would lead to a higher rate of photosynthesis unless the other factors promoting biological growth were also favourable. Thus the sink for CO_2 is the ocean.

The warm surfaces of the oceans are in equilibrium with the atmosphere, and would respond to increased CO_2 by taking more of it into solution – about a tenth of the increase. Its penetration into the deeper layers would be much slower, and the response time of the deep ocean to a change above may well be of the order of a thousand years. Broecker and his colleagues consider that at any time 60 ± 10 per cent of the cumulative increase in CO_2 will be in the atmosphere.

From the source already quoted, it seems that the cumulative total of CO_2 produced by burning fossil fuels was about 30×10^{10} tonnes by the end of 1960, two-thirds of which occurred after 1910. Converting the units from the table, one finds this amount to be around 15 per cent of the total quantity in the atmosphere naturally (if that is the right word) and again following the argument on dispersion out of the atmosphere, leads to the conclusion that the atmospheric increase at that time was 8 per cent.

* Donald W. Hood (ed.), *Impingement of Man on the Oceans*, Wiley-Interscience (U.S.).

Assuming an annual increase of 4·5 per cent in the tonnage of fuels burned, by the year 2010 production of CO_2 would be four times the present rate. The cumulative total at that time would be about seven times greater than the 1960 figure, and would equal about 210×10^{10} tonnes, or about the same as the natural quantity of CO_2 in the atmosphere. Following the reasoning given above, about 60 per cent of this quantity would actually be in the atmosphere. Broecker thus concludes that the quantity of CO_2 in the atmosphere will double some time in the first half of the next century.

Whether or not the rate of increase of burning will be maintained until 2010 and beyond can best be guessed from the discussion on the world's future demands for energy given in Chapter 13. An increase in climatic temperature would tend to reduce burning in temperate climates because domestic heating would be reduced, but in hot climates the air-conditioning load would increase, so having the opposite effect. But there are many other imponderables: fossil fuels are not the only past factor; what is the effect of glacial climate on the argument (CO_2 is highly soluble in icy water); and above all what are the major causative forces in climatic change?

In contrast to the predictions just discussed, it is worth quoting from the First Annual Report on Environmental Quality (August 1970) of the Executive Office of the President (U.S.A.) which points out that increased cloudiness and 'particulate suspension' in the atmosphere could lead to a lowering of temperature. An increase in average cloudiness from the 31 per cent measured to 32 per cent could cause a drop in mean global temperature of 0·8 deg.C. Estimates of growth of CO_2 in the atmosphere, published in 1965, are less than those discussed above, largely, it would seem, because they assume that two-thirds of the increase will be absorbed. The statement in Chapter 2 that the ocean is a heat store with a buffering effect on changes in temperature flatly contradicts all the predictions noted here.

Radioactive wastes

Nuclear power stations have the same problems of waste heat as coal- or oil-fired stations but they have no chimney, make no CO_2, sulphur, ash nor soot. But radioactivity as inherent in the fuel and transmitted to materials in close contact with the fuel, has to be contained. There are also radioactive wastes from the processing of nuclear fuels; such wastes are a substantial part of the problem for present nuclear power stations, and for fast reactors too, which will generate similar wastes as well as some that will persist for a quarter of a million years.

Radioactive waste materials that decay quickly present no great problem; they can be stored safely within the confines of the power station or processing plant until they are no longer active. Those that persist – in practical terms for ever – have to be interred in a coffin which is a good radioactive shield and will remain so, indefinitely. Practice has been to dump such coffins in the deep ocean. There has been much outcry and alarm at the spread of this practice and it is now controlled internationally by law. If a coffin leaked, it could spread radioactivity far and wide across the sea, but even that prospect is not so bad as the dumping at sea by the U.S. army of their stock of highly toxic nerve gas. Poison on land tends to remain more localized, even if the method of isolating it breaks down.

Graveyards of radioactive material are an obligatory product of nuclear power. The reactors themselves, when the power station has ended its useful life, will become great indestructible monuments with radioactive interiors.

The official attitude to radioactive wastes in the British nuclear industry was summed up recently in some parliamentary replies in these terms:

> All these radioactive wastes are disposed of in accordance with authorizations granted by Ministers. Gaseous discharges, after treatment where necessary, are released to the atmosphere. Low-activity liquid wastes are normally discharged direct to water courses or the sea. In both cases considerable dilution occurs after release to the environment. Depending on their nature and activity, solid wastes are stored, disposed of by burial on special sites which are monitored, or dumped at sea under internationally agreed conditions. . . . The 1960 Act ensures that disposals of radioactive waste neither add significantly to radioactivity in the general environment nor give rise to a public health hazard . . .
>
> No member of the public in this country has suffered injury as the result of gaseous discharges into the atmosphere or the discharge of low-activity wastes into water courses . . . nuclear sites have been required to report any occurrence involving emission of ionizing radiations or the release of radioactive or toxic substances likely to cause death or serious injury. No such occurrences have been confirmed. Minor releases are known to have taken place from time to time but apart from the incident at Windscale [Chapter 10] in 1957 there has been no significant emission of radioactivity.*

* Eldon Griffiths, Parliamentary Under-Secretary for the Environment, 20 February 1973 and P. F. H. Emery, Parliamentary Under-Secretary for Industry, 1 March 1973.

Oil in the sea

Slicks have been observed in the sea from natural seepages of oil, but they must be rare nowadays, alongside those left by man. Slicks are not permanent. In sea water, oil is broken down by biochemical processes, just as sewage and wastes in streams are broken down biochemically, so that within limits purification takes place as part of a natural cycle. If the oil is emulsified – in nature by mixing through the action of wind, wave and tide – its degradation is accelerated. The harm it does – by killing sea birds and fouling beaches – is reduced or eliminated if it can be emulsified. It eventually becomes food for fish.

There is no reason why every river and stream, even in industrial, crowded places should not be pure enough for fish to live in, if the local people wish it to be so. Pollution in the sea from oil could, similarly, be eradicated. There is complication because of the international nature of the problem, and emergency action would always be needed to deal with a tanker in distress. Pollution control rests on study and understanding of the practical operation of oil tankers – a complex subject – from which effective regulations would stem, together with research into the behaviour of sea-borne oil and into the biochemistry of its degradation, so that the most effective methods of dealing with it can be worked out.

The rapid increase in consumption of oil throughout the world, most of it transported by sea, and the use of tankers of enormous size (Plate 5) – half a million tons today, which is ten times larger than fifteen or twenty years ago – accentuate the scale of possible pollution, but may also force forward the application of effective remedies, which, according to many commentators, are today sadly lacking. Two recent events involving pollution of the sea with oil are noted in Chapter 10.

Heat as pollution

About two-thirds of the heat generated by burning fuel at a power station is rejected, either to the atmosphere through cooling towers, or to a river, an estuary or the sea. Enormous quantities of water are circulated between the condensers of the power station, where the water picks up heat, and the cooling towers or outfall, where the water itself is cooled. It is fashionable to call the rejection of heat thermal pollution, but whether it is quite unimportant, a minor blessing or a minor curse, seems to depend on circumstances.

Cooling towers are frequently capped with clouds which may descend to ground level and envelop roads and buildings. They can be dense and wet, and if persistent, seem to promote the growth of green slime. Local heating of an estuary can change its ecology, and the species of marine life that flourish. The same thing happens in a river, but the scale of it can be different. The classic example is the Trent, which exists by courtesy of the C.E.G.B. The heat rejected by the power stations sited along the Trent could quite happily evaporate the whole river, but its top temperature is kept to a certain limit, and it is used as make-up water to replace the dew-laden clouds which swirl away from the platoons of gigantic cooling towers which are deployed down the valley. The main pollution of the Trent is not heat, but sewage. It could be a major source of water for Britain, but it cannot be exploited until the sewage which is discharged so plentifully into it is first purified.

Thermal pollution, as this phrase is used in the precious sense just defined, is also held up as a poisoner of the whole world – which is how I translate 'polluting the ecosystems of the biosphere' – by some critics. It is illogical to blame waste heat alone for such a crime, for all energy, whatever its form or use, is eventually degraded into heat which is dissipated to its surroundings. To assess this problem, one has to look at worldwide thermodynamics and compare human energy usage with natural phenomena and with the earth's thermodynamic balance.

The theory is that the world's natural balance sheet of energy is made up of income from the sun, spent by re-radiation off the earth, with the escape of the earth's geothermal heat as another factor. The terrestrial heat engine – to repeat a term used earlier – distributes and rearranges, but overall there is a balance that maintains constant the average global temperature. That is the balance sheet; fossil fuels are on capital account and burning fossil fuels or generating nuclear energy add new sources to it. Present opinion is that if the influx of 'capital' reaches 5 per cent or so of 'income', then the balance will begin to change and the overall mean temperature of the earth's surface will increase. This is a pretty rough assessment as I think will be clear from what has gone before, but it is this line of thought which is really at the root of the idea of what is called thermal pollution.

The fringe of the subject

There is a distinction between a waste product and pollution. Waste products, notably spent and residual radioactive substances and ash and clinker from conventional fuels, become pollutants only if they are dispersed by accident or carelessness. They may, however, give rise to unsightly spoil tips, or scars on the landscape, and indeed the power station itself may be condemned as a wound if not a scar, but these are wider subjects than pollution and belong in later chapters.

To sum up

It is reasonable to take a cautious, but qualified, optimistic view of the future dirt and destruction caused by energy usage, even with the exponential growth that lies ahead. Problems that will have to be solved, taking only the next doubling on the world scale by the 1980s, are severe. The lessons are that care and dedication by countless people are essential; that the economic drive for cheap energy is powerful and necessary, but is not itself enough; and that balanced attention must be given to all the ramifications of energy usage, within a framework of reasonable social and political stability. In terms of pollution, these standards include vigilance in observing and correcting new effects: for example has London's air deteriorated more from the motor vehicle than it has gained from the Clean Air Act?

Perhaps I can quote a second time from Mr McNamara's speech at Stockholm. He was concerned primarily with the World Bank's projects in the developing countries, and he said:

> What we have discovered is significant. By careful analysis we have found, in every instance to date, that we can reduce the danger of environmental hazards either at no cost to the project, or at a cost so moderate that the borrower has been fully agreeable to accepting the necessary safeguards. Central to the success of this approach is the principle that in the issue of environmental damage, prevention is infinitely to be preferred to cure. Not only is it more effective, but it is clearly less expensive.
>
> Responsible officials in the developing countries are aware of this. We in the Bank have found no evidence that they are unresponsive to what can be demonstrated to be a serious ecological hazard or a threat to health or social well-being. It is unfair to suggest that the poor countries are indifferent to the environmental issue, and simply dismiss it out of hand as a rich nation's problem. They do not.
>
> It is not surprising that those who call for a slowing down or complete halt to economic growth tend to be those who are already amply provided with the advantages which that very growth has made possible.
>
> What I mean by the environmental advantage of the late-comers to the development process is that they can far more easily and inexpensively build into their industrial infrastructures the practical preventative measures necessary to avoid the ecological damage which the developed world has already suffered. Our experience in the Bank confirms this. There is an increasingly broad variety of

anti-pollution technologies available to the poorer countries – technologies the affluent countries have had to develop at a far later and more difficult stage of their industrial expansion.

Those technologies can work and work well.

In a word, clean housekeeping on the industrial and national scale is very broadly practical, possible and above all essential.

Do it yourself

So the comfort, the dashing about, and the getting rich of modern man needs energy, and generating that energy throws off a heroic spume of unwanted muck and wastes. Most of it can, or could, be contained or cleaned up at rather modest cost in relation to the even more heroic wealth of the useful fraction. Some progress, but often not enough, is being made. On the other hand some critics exaggerate, and draw unnecessarily alarmist conclusions.

Since man himself is a heat engine, these effects are a magnification and a diversification of his own metabolic processes. Bernard Shaw, with an unnecessary injection of class-consciousness, otherwise accurately described the rich as inefficient machines for turning good food into poor fertilizer. Starting from the austerity of scientific logic rather than the extravagance of literary posture, Hugh Nicol* related energy to life by arguing, amongst other themes, that food, especially protein, was a special class of fuel, in short supply and of special importance: 'External energy is not enough. Unless translatable and translated into terms of food chemistry, the possibilities of energy are only potential and secondary for purposes of living.' Food as a fuel and sewage as a pollutant come within the ambit of our theme, but I will spare the reader a treatise on agriculture on the one hand or sewage purification on the other, and be content to ornament this particular approach with two little digressions.

First, human organisms reject energy which is still high-grade enough to be utilized in a heat engine. Fig. 12 shows an ingenious example of energy conservation. Sewage sludge has to be heated to digest (i.e. purify) it, and it gives off methane in the process. This methane is used as fuel for an engine; the cooling water from the engine is itself cooled by circulating it around the sludge, which needs to be heated to maintain digestion. This is probably the nearest thing in engineering to perpetual motion. Unhappily, new trends in sewage purification are likely to make the practice obsolescent.

Second, Hugh Nicol's point is likely to be overtaken by events.

* Hugh Nicol, *Microbes and Us*, Pelican Books.

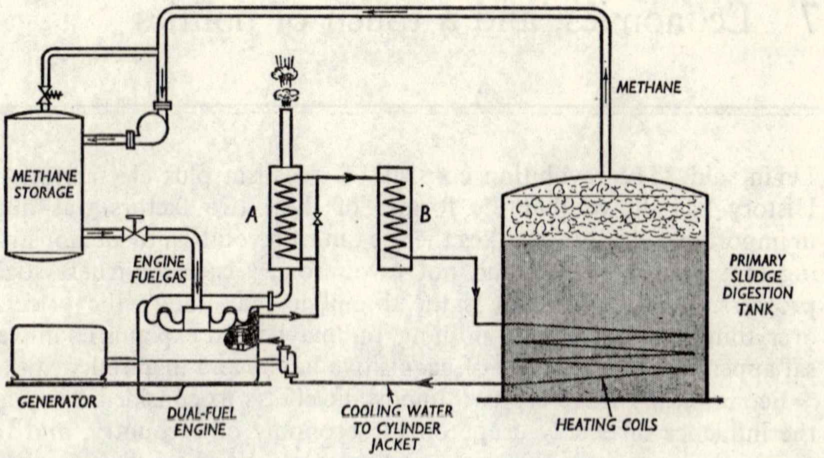

Fig. 12 The cycle of heat exchange between a dual-fuel engine and a primary sludge digestion tank. Hot water from the cylinder jacket is heated still more by passing it round the engine exhaust in the heat exchanger A. As the summer approaches, not so much heat is needed to digest the sludge, so the heat exchanger A is bypassed. Then in the hot weather still less heat may be needed for the sludge tank, so the heat exchanger B is brought in, to cool down the circulating water. The thermal efficiency of the cycle can be over 80 per cent at full load, with a modern installation.

Research may soon succeed in making edible protein from crude oil – from crude to food, if I may anticipate a television commercial of the eighties – and apparently a fraction of what is now spilled at sea would make enough protein for the even more populous humanity of the years ahead. Indeed thermal pollution rather than protein could one day be its ultimate problem according to Dr Fremlin,* who draws a delicate picture of a world population of 10^{16}–10^{18} a thousand years from now, the problem of supporting which appears to be primarily thermodynamic, requiring the earth to be roofed over with a black outer skin at a temperature of about 300 deg.C (perhaps even 1,000 deg.C) to get rid of the waste heat generated by the population.

* Dr J. H. Fremlin, 'How many people can the world support?', *New Scientist*, 29 October 1964.

7 Economics, and a touch of politics

Lenin said, 'The revolution consists of socialism plus electrification.' History has shown that the former of these two factors was quite unimportant; economics takes the fire out of revolution by demonstrating that price per kWh and not revolutionary excess furthers social progress. True, economics is the discipline that knows the price of everything and the value of nothing, but maybe that explains its universal appeal in this great age of acquisitive habits and material wealth.

Economically there are two important effects to consider. They are the influence of energy usage on the economy of a country, and the money and resources that have to be deployed to have that energy – that is, whether it is cheap or dear. Philosophically, there are two other effects of special interest. Electricity supply lends itself to erudite analysis of economic factors; much enquiry and research and new ideas in economic theory have spread from the experience of electricity supply. Electricity is to economics rather like heat engines have been to science. The economic aspects of other forms of energy tend to be less academic than those relating to electricity supply, and while important, generally devoid of subtlety. Lastly, one notes the utter unimportance, in economic measurements as they are valued today, of conserving energy. The cost of the useful fraction is the yardstick, and reduction of losses has to be justified by savings in cost.

The economics of 'work done'

In a well-known analysis of urban roads and traffic,* the generation of traffic was discussed, and the notion put forward that traffic is a function of activities. If people are busy doing this and that, then traffic starts to flow; to stop it stops or hinders the activities. A fact so strikingly obvious needed a genius to state it. Energy is a close analogy. If a community wants to do jobs of useful work, or to have standards of comfort and convenience that require the consumption of energy, then the economy of that community will be retarded, and maybe heavily penalized, if energy is not available in sufficient quantity when and where it is needed. But an over-generous supply of energy will not alone stimulate economic growth. The economic optimum lies in predicting and meeting the demand.

This important principle can be illustrated by many examples, from

* *Traffic in Towns* (the Buchanan report), H.M.S.O., 1963.

the lack of applications for early electric motors in the nineteenth century, and the small demand for oil in Canada and the U.S.A. when the first oil wells were introduced, to an abundance of modern cases. Some of these modern examples are in the developing world – that is, those countries that do not yet have an industrial economy. One well-documented example is the Owen Falls hydro-electric scheme in Uganda, which it is instructive to select for comment.

The scheme was completed in 1954; it impounds the Victoria Nile and controls the level of Lake Victoria over a small range, thus providing what is called century storage on the Nile for the benefit of Egypt, which is a novel long-term feature and outside the scope of this discussion. From an engineering point of view, the scheme is favourably sited, and is in all respects an accomplishment of high standard. It has an installed capacity of 150 MW, but only 30 MW was installed initially.

In the post-1945 period of African colonialism, there was a strong feeling that the benefits of modern industry and production should be extended to the materially poor people of that continent. Provision of electrical power was considered to be a fundamental requirement (Lenin must have been on one of the Colonial Office committees). Owen Falls dam was built as 'an act of faith' with the idea that if cheap power were available, then industry would be attracted and economic acceleration would follow. Unfortunately it did not. Demand grew so slowly that less than half the scheme's possible output was being generated after twelve years, and one-third of that amount was being exported to Kenya.

A fat financial return must be expected in the early years of a scheme's life if it is to pay off its capital cost, and some very tangible economic benefit must have been created to justify its construction. If not, then the resources devoted to the scheme would have been better employed on some other scheme which gives a better return.

One of the great tragedies of civil engineering is to see a reservoir with a magnificent new dam, that will never fill with water. One such I saw in Turkey, where about thirty sink holes had formed in the karstic rock of the reservoir, and had turned it into a colander. Karst is a magnesian limestone, and forms solution cavities; it is an engineer's nightmare. The failure in such a case is brutally self-evident. But an economic failure, where there is no demand beyond the substations at the end of the high voltage line, is equally futile, even if it is not placed directly in front of one's eyes. It has one mitigating factor however. Time brings growth, and the economic error is likely to be rectified if one is patient enough to wait. At Owen Falls, there was no competition for its capital funds, which would probably not have been used at all if the scheme had not been built. Ergo, it was the right economic answer? Not by today's standards of assessing economic development, and the critical need for capital to finance it, in the world's developing countries.

Prestige projects, amongst which hydro-electric projects seem to be rather favourites, can be white elephants rather than economic accelerators, an experience which has been learned at great cost in several developing countries. The practical approach is to forecast the demand. For electricity, the future activities of all classes of consumers are surveyed, and how much energy they will need is estimated, by area or by group or by various other detailed techniques, over a period of years ahead. Then the least-cost method of meeting that demand is devised.

That the supply of energy does not itself accelerate economic growth is now widely accepted. The reasons why are explained in this quotation from a lecture given by one of America's leading authorities on the subject, Philip Sporn.*

> Neither the absence of energy nor the presence of high-cost energy is a necessary cause of low income; nor are they a bar to high income. On the other hand, although the availability of abundant low-cost energy supplies may be a factor in promoting economic growth, such growth is not a necessary result. The availability of energy resources in their natural condition in the ground, or even the actual production of energy, is not enough. Energy consumption is the significant factor, and consumption need not be related to production in any particular area or region.
>
> The development of nuclear power has focused special attention on electric energy, but here again, since electric energy is energy in a refined form, the same basic principles we have been describing with respect to energy in general also apply. But in the case of electric energy it can be seen even more clearly. Unlike the primary energy sources electric energy cannot be produced unless it is consumed. Indeed, although the elapsed time is so brief as to seem instantaneous, in actual fact consumption must precede production; that is, a load must be present which sets in motion the complex equipment that will deliver the electric energy called for by that load. Thus, oil, coal or gas can be produced and stored; electric energy cannot be produced and stored except in small quantities in storage batteries. Electricity must be ready for use instantaneously with the speed of light as it is called for by machines, tools, light fixtures, and other consuming devices.

* 'Energy: its production, conversion and use in the service of man.' Three lectures at Columbia University. Commonwealth and International Library of Science and Technology, Engineering and Liberal Studies. Pergamon Press.

Once this is recognized, we are in a better position to analyse some of the proposals to promote the development of human society by making available electric power facilities.

In a modern industrial society, electric energy use is pervasive. In an inter-industry relations study for 1947 by the U.S. Bureau of Labor Statistics, the country's economy was classified into 190 component industries; these included manufacturing, mining, agriculture, commerce and services. All but three of these 190 industries consumed significant quantities of electric power; the only exceptions were cotton farming, oil bearing crops and fisheries, and hunting and trapping. Today more than 98 per cent of the homes and farms in the United States are wired for electricity, and their consumption of electric energy in a great variety of common appliances is growing at a rapid rate. The pervasiveness and rising levels of consumption of electric energy in the United States is a measure of the growing productivity of the economy of this country. It is also a measure of the increasing conveniences and aids for comfort and ease of living in the home.

But electric energy, as I have already emphasized, is only one among many factors important in the production of material wealth. If you examine the data on the cost of electric energy consumption for the 20 manufacturing groups as classified, for example, by the U.S. Census of Manufactures, you will find that electric energy costs are less than 2 per cent – a minor percentage – of the total value added by all manufacturing. If you compare the cost with the value of product shipped, the figure would be reduced well below 1 per cent.

This is true of manufacturing in general. It is not true of those few industries where electric energy enters into production almost as a raw material. In those cases you will find the cost of power is more significant. Thus for such industries as aluminium reduction, or the production of zinc, chlorine, and similar chemical, electrochemical, or electrometallurgical products, cost of energy may represent as much as 10 per cent or 20 per cent of the value added by manufacture.

Energy is essential to operate the capital equipment which, when applied to production processes, increases productivity. The need for energy results from the marshalling and use of capital equipment in production. The availability of capital, therefore, in the form of tools, machinery, and equipment, is a fundamental requirement for increasing the consumption of energy to raise productivity. An existing or potential supply of low-cost energy may encourage the use of energy-consuming capital equipment. It may catalyse other forces necessary to bring about industrialization, but this is not an inevitable consequence. A flourishing industrial society requires many highly developed human and material resources, of which energy is only one.

The only categorical statement that can be made is that energy, particularly electric energy, is critically important if absent. But I think I have shown that, if other factors are favourable, energy can be made available with relative ease in almost every part of the world – at least under present conditions in the world's resources.

Among the numerous factors that determine the extent to which machinery will be applied to production, availability of capital is crucial. This is because the cost of electric-energy-consuming equipment in almost all cases far exceeds the cost both of the electric energy required for operation of the equipment and the capital cost of the electric-energy-producing facilities. This is equally true of an electric shaver, that has an annual power consumption of 1 kWh with a cost that might range from 1 cent to 3 cents, as it is of a gaseous diffusion plant which might have an annual consumption of almost 20 billion kWh, but which also represents an investment of about $1 billion.

The cost of growth

One can conclude from Dr Sporn's thesis that Lenin was right about electrification, but by accident. However, the other factors which are necessary complements have all been active, and, universally in the industrialized countries of the world demand for electrical power has grown constantly from the time that electricity was introduced. The industry plans well ahead, detailed planning extending say for seven years, which is about one year longer than it takes to design and build a power station. Great sums of money are invested, constituting a sizeable proportion of a country's overall capital programme.

As we have seen, considerable success in reducing capital cost has been achieved with conventional thermal plant, which has reacted against the popularity of nuclear and hydro-electric power stations, unless their costs too can be brought down to similar standards. Nuclear and hydro power have high capital costs and low running costs. Conventional thermal is about half and half, so the price of money – the interest rate – has a bigger effect on the first two. In recent years they have had to face increased charges for amortization as well as the purely technical competition. Apart from the power stations there are other capital costs, usually less but of the same order of size, for the other parts of the supply system – transmission, distribution and the connection of consumers.

Evaluation of cost per kWh 'sent out' from the power stations is complicated by the peak load. Not only has the life of the power station to be considered, but also its load factor. The nature of the various installations in the system determines the economic problem as explained in the passages on peak load and operation in Chapter 4. Peaks may be sharp, and last for an hour or so, or be longer and less acute. Different methods of meeting them will be devised to get the best out of the system. 'Spinning reserve' has to be kept in the system ready to go on load as the peak builds up. Boilers have to be banked, and hence fuel used extravagantly, if coal-fired stations serve for peak duty. Hydro stations are more flexible if they have storage, and are preferred. Gas turbines are useful for peak and standby. Pumped storage has a useful role.

There is considerable nicety of analysis in estimating maximum demand and the shape of the demand curve in the future, and then evaluating these various methods and selecting from them proposals which are the least-cost method of meeting the predictions with a given margin of stand-by capability. The answer can turn on fine points, both of economic analysis and of physical performance. I have heard it said that a blunt pencil used to plot a graph can transfer a few million pounds from one side of the balance to the other. More seriously, the analysis may lead to an unusual specification for achieving a least-cost solution, and may be the spur for an original engineering design.

Flattening the load curve

A complementary method of attacking the peak load problem is to try to flatten the load curve, either by the persuasion of differential tariffs or by mandatory disconnection. Economists try to evaluate the economic costs to the community of shedding load, but whatever their answer, arbitrary disconnection is disliked by any electricity utility that has pride in its competence. Off-peak tariffs are widely and successfully applied in industry. The most ambitious attempt to apply them to domestic consumers has been the British off-peak tariff, coupled with the sale of storage heaters, which use energy at night.

The size of the investment that goes to meeting the peak demand of a national electricity supply system is of justifiable concern to economists, and, to repeat a point made in Chapter 4, there would seem to be opportunities for economy by devising tariffs which give incentives to consumers just sufficient to flatten the demand curve. The economics of tariff structures seems to be a neglected subject.

Economic issues of heat–electric generation

Some wider economic issues are thrown sharply into relief by discussion of heat–electric generation. The British system has as its aim cheap electricity, and economic analysis serves that aim. The economic picture has different shading if either a place is found in the load curve for combined heat–electric sets as suggested in Chapter 4; or if electricity is considered as the byproduct and heat the principal product; or again, if the older thermal sets which are restricted to peak or middle demand are downgraded electrically by working them at higher exhaust pressures, but using the exhaust steam as a source of heat for district heating. This last method could be applied to take some of the place found for combined heat–electric generation in the load curve, and has potential economic advantage because it would extend the lives of the turbines so modified after they had been written off in accounting terms, and also because, in Britain, the shape of the load curve, with its 6:1 variation between maximum and minimum demand, is an economic handicap and restricts utilization of the middlegrade generating sets.

If cheap heat were available from district heating, the domestic electrical load for heating would be reduced, with some easement in the increase in demand, as well as the conservationist advantage of using fuel at higher thermal efficiency. In the early days of using electricity for heating, it was a common point of discussion that high-grade energy should not be used for low-grade purposes, but this principle seems to have faded away, and cost dominates the argument. But as the exponents of 'pure' electricity point out, it is not correct to compare their 35 per cent thermal efficiency with the 80 or even 85 per cent of the combined system, since the heat load would in any case be met at high efficiency, and if heat and power loads are not balanced, thermal efficiency drops.

How to evaluate these factors in terms of national economic costs is an interesting exercise, which can be recommended to economic researchers. One precaution has to be suggested. The technical characteristics that are available, or that are possible, or that can be designed for, have a profound influence on the economic answer, and deep knowledge of both disciplines (including on the engineering side a basis of practical operational experience) has to be carefully integrated.

I can recall an investigation of this kind into the peak load and tariff problems of a country that relies on hydro-electric power and has a big electrical pumping load for irrigation. A drought increases the load, and depletes the source of energy as well. A voluminous and detailed economic study set out to ascertain how to adjust the tariffs to obtain the greatest overall national benefit. It concluded that the greatest economic advantage would follow if the irrigation pumps were operated only when it was raining! A process called inter-disciplinary co-ordination was initiated to modify the presentation of such an outstanding fact.

How to relate, in terms of national economic cost and benefit, a reduction in demand for say, electricity, combined with increased use of resources for district heating should also be a happy hunting ground for economic research. The same fuel for both could be assumed; if higher-grade energy were saved for high-grade uses, thermal efficiency improved, and cheap heat supplied, then the electrical grid would be used less, and extra economic costs incurred in installing steam mains etc. I draw attention to this because there is need for guidance, in broad economic terms, when choosing between alternative methods of carrying out the same task. In these terms – which are logically, if not practically and politically, the only valid ones – a national policy for fuel and energy does not exist.

The practical contrast is that all methods are available to the customer, and he chooses on the basis of price – probably inadequately defined, unless he has professional advice – personal preference, or the abilities of the salesman. Sadly, in Britain district heating has to be excepted from that statement for most places.

Returning to the remarks on economic evaluation, whatever results emerged from any such study it can be guessed that greater economic benefit could be achieved more simply by attaining higher standards of everyday heating practice and of thermal insulation. There is also another economic balance to be weighed, between cheap heat used extravagantly, and expensive (i.e. expensive to the consumer) heat used sparingly with extra expense for thermal insulation.

Price per kWh and social welfare

Electricity utilities may be publicly owned, or private companies. In both cases their optimum aim, almost universally, can be shown to be to supply power at the most competitive price whilst remaining in a healthy financial condition. In practice such a utility should be run so that it has revenue sufficient to meet its operating costs; to allow for depreciation of its assets; to repay its debts (consisting of capital sums borrowed to pay for its new installations), and also to make a reasonable contribution towards the costs which will be incurred in the future for expansion and modernization of its supply system. When this policy is followed successfully, a lean financial healthiness results and the utility is well prepared to face the future; the best economic answer, in terms of national interest will in most cases have been achieved. If the utility is subsidized it tends to get flabby, and electricity supply seems to be one of the industries least able to ask in reasonable terms for such support, in view of its successful record and the nature of the service it gives.

There are exceptions to this general rule. The one most generally encountered is the need to make some sort of subsidy to encourage electrification of rural areas where the population is sparse and poor.

One interesting variation from the rule, which is unusual if not unique amongst modern electricity utilities, is the North of Scotland Hydro-Electric Board. There is a 'social' clause in the Act which gave authority for the formation and operation of the Board. The reason for this clause was that the Board would be exploiting a profitable natural resource and making its best profits by exporting that resource to the lowland industrial regions. The Highlands was a region that was economically backward, and hence had only modest need for the power; there was a history of local objection to building hydro-electric schemes.

So the Board gave preference to Scottish firms in its contracts, provided they were competitive, promoted the use of electricity in the Highlands, encouraged new industry in the Highlands, and built uneconomic distribution schemes to serve sparsely populated areas. In the event, it also made profits. Another feature of its policy was that it paid special attention to amenity, a word which more correctly describes what is now fashionably dubbed the environment. It buried its pipelines and power stations so that no one in the glen can see them; it built fish passes and salmon hatcheries to retain or increase the population of salmon in the rivers which yielded power, and it followed a generous policy of compensation for the owners of property and land affected by its schemes. This policy was at its height during the 1950s and 60s when most of the Board's schemes were constructed. Today it is building only a pumped storage scheme and a thermal station. A new economic surge is being felt in the Highlands from the discovery of oil in the North Sea, and it would appear that the Board's twenty years or so of maximum spending contributed substantially to tiding the Highlands over a lean period, without impairing the competitiveness of its product – surely an economic object lesson. During those years there had to be some stimulus to counter depopulation; tourism and forestry, the other main activities, would not have been enough without the Hydro Board.

This example also shows that the purposes and ramifications of schemes of river development tend to become rather wide, and to bring in many issues. Other examples are discussed in Chapter 9. Where simply the supply of electricity is the requirement, special cases are rare. Price per kWh then is the important yardstick and, remembering Dr Sporn's analysis, there should be very cogent reasons for departing from it.

National energy policy

Economic factors other than price per kWh do emerge, however, in a national context and where more than one fuel or method of conversion is in question. The discussion on heat–electric generation is a good example. Even if no clear-cut answer is obvious, there is clearly advantage in discussing and examining the alternative possibilities, and investigating whether one form of energy can be balanced against another, in order to promote the most economic national pattern of energy usage. The policies of a single-minded body such as the (British) Electricity Council could then be at odds with wider national interest, either in relation to a competing form of energy, or in relation to some strategic national aim. There are thus advocates of a national energy council which would assess and plan strategy. I have held over discussion of this topic until the last chapter, where it fits in with a little crystal gazing into the future.

National aims in most countries have up to now been rather crudely assessed, but this has not prevented governments from 'interfering' for all manner of reasons, some excellent, some disastrous. Lenin and his quotation are still with us. In Britain recently, for example, the C.E.G.B. has complained that it has been forced to burn coal instead of oil, and therefore the Government should pay the difference in cost (nationalized industries in Britain have to meet financial targets) or it should burn the fuel of its choice. Government control can vary from the complete and comprehensive, including supervising day-to-day management, to a few sweeping dictats which ensure that the operations of the energy industries comply with its strategic aims and purposes. The degree of government control tends to increase and to become more complicated; this trend is worldwide and there are compelling reasons for it.

The importance of energy in a country's balance of payments is one consideration. The extremes are the little sheikhdom of Abu Dhabi which has the highest *per capita* income in the world, earned from royalties on oil, and probably Japan which imports over 80 per cent of its energy. Another is the security of a nation's sources in times of political trouble or war, a reason which can lead to the protection of indigenous industries, such as the coal industry in Britain. This reasoning is reinforced if unemployment would be caused by running down the indigenous industry, and reinforced again if its interests are politically influential. In Britain coal is too dear; petrol is heavily taxed. (See Postscript. Events have overtaken the next paragraph.) In contrast to the complex economic calculations of electricity supply, it is a simple matter to figure whether oil or coal is the cheaper fuel, and to slap on a tax to make up the difference.

The oil industry is a unique phenomenon in this setting. It consists of a few powerful international companies (most of them effectively U.S. owned, but equally effectively international in their operations) which wield great power and influence and dominate, not only their own industry, but the pattern of energy usage throughout the world. It has been oil which has been exploited faster and faster to meet the world's growing demands. Road transport depends virtually 100 per cent on oil, and the oil industry is inflexibly linked with that use, but it has expanded greatly into many other fields. Natural resources have been within twenty or thirty years of exhaustion for the last thirty years or so, according to the strict numerical total of proven reserves, but reserves continue to be proven. Great advances have been made in all the technical aspects of extracting, transporting and refining oil, so keeping its cost down in face of extra difficulties, such as the closure of the Suez Canal. The dominance of these great companies however, seems to be coming to an end in the producing countries. A body called OPEC – the Organization of Petroleum Exporting Countries – has negotiated 'participation agreements' with the oil companies, which give the producing country a stake of 25 per cent in its own oil industry in 1973, rising eventually to 51 per cent in 1982. The oil-exporting countries will be expecting, with increasing control, an increased share of the wealth from their oil. Prices are expected to rise, but it will be interesting to see whether development and innovation will again be able to reduce that tendency.

All these arguments boil down to a few simple facts or principles. For the consumer, the size of his bill and the convenience and reliability of his service are paramount; the rest are talking points. For the country there tend to be simple alternatives. Dearer fuel – coal for example – at home is traded against the advantage in the balance of payments, maybe together with an easing of unemployment. Taxation is the biggest influence which a government can exert, but it must be used with some discretion because of the economic value of cheap energy in keeping a country's products competitive. There must also be a broadly acceptable tradition of policy towards planning and pollution.

Apart from a climate of growth, other changes occur rapidly and dramatically, like those in the British gas industry in the past decade or so. Gas was made from coal, and was a secondary source of energy. Gas-making processes from oil were introduced, and those from coal improved. Then natural gas from Africa was liquefied at its source, and imported into Britain in liquid form. All these developments were virtually abandoned with the sudden discovery and substantial exploitation of natural gas from beneath the North Sea. Gas is now one of

Britain's four primary sources of indigenous energy – coal, oil, gas and nuclear. It is not many years ago when there was only one – coal. Other changes have been going on as well as those in the gas industry.

There is a school of thought which pours cold water on the idea of an energy council. The various energy industries are best left on their own, it is argued. They are flexible and react to events readily as they occur, because they have to react to market forces and are stimulated by competition to improve their performance. In any case, changes occur unexpectedly and unpredictably, which would vitiate the grand plan of an energy council; the council's influence would then be useless or reactionary.

8 Let's give it a bit of environment*

The structures of power at their best have form, beauty and a style of their own. The architecture of power, also, has its power. Before giganticism set in, power stations were sometimes poked in a back alley and treated as shed-like structures of no consequence. The wooden cooling towers of that vintage seem to have the very spirit of shabbiness. But in other cases, where electricity was hailed as a bright new achievement, monumentality came with it, and pretentious power station buildings, some of them of architectural merit, were built.

The exponential growth of power has borne heavily on the sensibilities of architects. The scale of power stations began to dwarf conventional balance and motifs, and they began to present visual problems of their own. Battersea power station in London (built in the 1930s) was probably the prototype of the fashion for 'cathedrals of power', which was characteristic of the power stations built in the years after the war in Britain. Their style made use of brickwork which was conceived and detailed to suit the grand scale of the building. But the buildings continued to grow, even beyond the scale which that approach could handle; although the style itself was beginning to be outmoded architecturally, it was the cost of brickwork which decisively killed it.

An approach relying on a less formal balance of mass and surface, and making use of light cladding which was cheaper on this scale than brickwork, superseded it. This style has been used with ingenuity and success, and is still in vogue today. All the ancillary structures are brought into the balance of the composition, which thus attempts to group the cooling towers, the chimneys, and so on, harmoniously with the main building. Here lie great problems, and the bulk of the former and the height of the latter can neither be hidden nor muted. There are acres of switchyards, a coal store measured almost by the square mile, and a concentration of transmission lines converging with relentless tentacularity to attack the order of the composition. Yet no small success has been achieved.

In the Trent valley – 'power station alley' – the landscape is flat, and from almost everywhere at least one power station can be seen on the horizon, and often two or three. As one gets within a couple of miles or so of one of them, it begins to dominate the scene. Nevertheless, many

* Instruction given to an architect by his client.

PLATE 1. Nuclear energy in Cumberland. On the left is Calder Hall, the world's first nuclear power station, with, on the right, the Windscale works. (*U.K. Atomic Energy Authority*)

PLATE 2. Wylfa nuclear power station in Anglesey, North Wales, is one of the largest of the magnox stations, generating 1,180 MW. The station was opened in 1972. (*C.E.G.B.*)

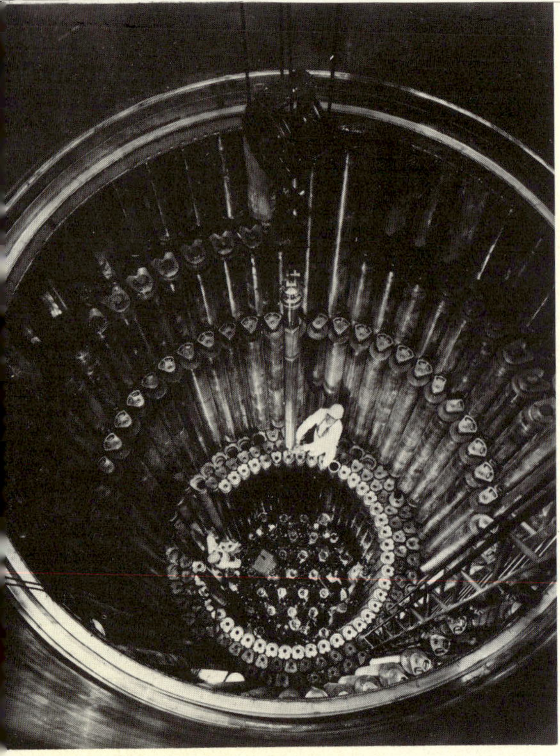

The core tank showing a dummy fuel-assembly being lowered into position.

Underside of the charge machine's rotating sh
This machine removes and replaces fuel rods w
the reactor is in use.

PLATE 3. The prototype fast reactor at Dounreay, Caithness, Scotland. (*U.K. Atomic Energy Authority*)

Intermediate heat exchanger being lowered into position.

The primary pump which circulates liquid sodi

PLATE 4. Megafilth efflux. Europe's largest chimney, the 850-foot giant at Drax power station in Yorkshire. (*C.E.G.B.*)

PLATE 5. *Globtik Tokyo*, 477,000 deadweight tons, is the world's largest ship. Built by Ishikawajima-Harima Heavy Industries Co. for Globtik Tankers of London it is chartered to the Tokyo Tanker Co. for hauling crude oil from the Persian–Arabian Gulf to the Nippon Oil Group's Central Terminal at Kure. (*Globtik Tankers*)

PLATE 6. The television transmitting mast at Crystal Palace, London, during smog. The flat upper surface of the smog is typical of the atmospheric inversion that causes it. The top of the mast is 1,074 feet above sea-level, and the mast itself is 712 feet high. The upper surface of the smog is approximately 700 feet above sea-level. (*C.E.G.B.*)

PLATE 7. Industrial smoke in New York City. (*Fairchild Aerial Surveys*)

PLATE 8. In the infamous London smog of 1952 power stations did not contribute to the pollution. This photograph proves the point; it shows the plume from Deptford power station's chimneys (left) penetrating the smog and being dispersed in the clear air above. (*C.E.G.B.*)

PLATE 9. Ski-jump spillway discharging at the Barrage de l'Aigle in the Massif Central in France. The twist in the jet is deliberate and helps to dissipate the energy of the overflowing water. The ski jump passes over the roof of the power station, part of which can be seen on the right. (*Électricité de France*)

PLATE 10. The Pieve di Cadore dam on the river of the same name in the Dolomites, shown here at the time of its completion in 1950.

PLATE 11. Channel of the ski-jump spillway at Genissiat on the Upper Rhône in France. (*Compagnie Nationale du Rhône*)

PLATE 12. The machine room at Genissiat. (*Compagnie Nationale du Rhône*)

PLATE 13. Cruachan pumped storage hydro-electric station in the Highlands of Scotland. This is an underground station; bare rock can be seen exposed at the end and at the side of the generating hall. (*North of Scotland Hydro-Electric Board*)

PLATE 14. Beauharnois hydro-electric power station on the St Lawrence river. The canal forms part of the St Lawrence seaway. (*Hydro Quebec*)

PLATE 15. Wheeler dam on the Tennessee river in Alabama. (*T.V.A.*)

PLATE 16. Cooling tower at Ferrybridge, Yorkshire, collapsing in a gale. (*C.E.G.B.*)

PLATE 17. Einstein enlisted. U.S. Marines spell out Einstein's mass–energy equation on the deck of the nuclear-powered carrier U.S.S. *Enterprise*. (*U.S. Navy*)

PLATE 18. The latest storage structure for oil, shown here before being floated out to its permanent position on the bed of the North Sea at Ekofisk Field. The perforated wall is to reduce the force of waves. Inside is a four-celled tank. (*Phillips Petroleum*)

PLATE 19. A diesel engine—thought to be the world's largest—with a maximum continuous rating of useful power of over 30 MW. Built under licence by Eriksberg Mek. Verstads AB of Gothenburg, it powers a 275,000-ton ore–oil carrier. (*Wabco Westinghouse*)

Plate 20. The intermediate pressure steam turbine line for one of the 550 MW turbine-generator sets at Thorpe Marsh power station near Doncaster, Yorkshire. The set is cross-compounded like the sets at Longannet described in Chapter 11. (*Reyrolle Parsons & C.E.G.B.*)

Plate 21. The runner of the Francis turbine for the El Chocon hydro-electric scheme in Argentina, assembled at the maker's works in Britain. (*Boving & Co. Ltd*)

PLATE 22. A view of the Aswan high dam under construction, 1964. The Nile has now been diverted and flows through the tunnels. Above the tunnels is the structure that now carries the control gates. The power station is at the other end of the tunnels. (*Central Press Photos Ltd*)

PLATE 23. The control room at Longannet power station in Scotland. (*South of Scotland Electricity Board*)

PLATE 24. Longannet: general view. A coal conveyor is in the foreground. The tall building contains the boilers, and to the left of it is the machine hall. (*South of Scotland Electricity Board*)

PLATE 25. Longannet: the machine hall. The four generating sets each have two shafts. In the foreground is a high pressure turbine; the intermediate pressure turbine is on the next shaft, fed with steam from the two overhead, lagged pipes. (*South of Scotland Electricity Board*)

PLATE 26. Longannet: the 275 kV switchgear, housed in a building next to the machine hall. (*South of Scotland Electricity Board*)

PLATE 27. Longannet: before being fed to the furnaces coal is crushed to the consistency of fine flour in these pulverizing mills in the basement of the main building. (*South of Scotland Electricity Board*)

PLATE 28. The Rance barrage in Northern France, the world's only sizeable source of tidal power. In the background, right, is the harbour of St Malo. (*Électricité de France*)

local people admire the power stations, and speak well of them architecturally. A Civic Trust award for 1968 to West Burton power station (*see* jacket photograph) brings out the point rather well. The assessor's comment is lengthy but fascinating. He criticizes various artifices of design which were clearly intended to beautify the station – as if, perhaps, one were putting cosmetics on a hippopotamus – but then goes on to express strong sentiments about the strength and character of this gigantic edifice and the flat landscape in which it stands, as follows:

> On the debit side, therefore, West Burton has strained to do too much: wind, weather and haze have conspired to defeat many of these efforts and we are left, happily, with an immense engineering work of great style which, far from detracting from the visual scene, acts as a magnet to the eye from many parts of the Trent Valley and from several miles away. The asymmetrical grouping of the cooling towers is most effective and their relationship with the main buildings is very satisfactory. Most encouraging in principle and successful in effect has been the planting, even miles away from the new station, of groups of trees in order that the building may look well when seen from a distance. The encouraging significance is that these trees have 'local' value in themselves and thus the massive investment of a generating station can be made to extend certain benefits for miles around, even though very few people are aware of the fact. West Burton may be the last of the big twin-chimney stations, which are being superseded by such huge single-stackers as Cottam and Ratcliffe-upon-Soar. While much can be said in favour of the compact masculinity of the most recent type, West Burton, with its undoubted chic, must really be the Beau Brummell of Nottinghamshire, and as such it fully deserves its award.

The waisted natural-draft cooling tower first became prominent in postwar Britain, and immigrated into the U.S.A. about twenty years later. Le Corbusier found its shape so exciting that he had to have one, and so he designed a church which consisted of the usual cruciform plan with a 'cooling tower' on the top of it, over the crossing in place of a tower or spire. Certainly a cool cathedral of power, but I will leave the reader grasping for the neatest *mot* which can link theology and thermodynamics.

A group of eight cooling towers is needed for each of the British 2,000 MW stations. At West Burton, an attempt was made to relieve the visual giganticism of the group by painting the towers in different colours, so that one or other stood out and the others were less accented, according to the time of day and the strength or absence of the sun. The idea is intriguing but I have not yet heard a kind word spoken in its favour.

Some other industries, for example the steel industry, have occasion to erect structures as massive and dominating as power stations. A similar style is often adopted, so power station architecture is not unique, but it does embody the best of contemporary style as can be expressed within such stark limitations of shape and scale. Nuclear stations differ from conventional thermal ones only in that the reactors add new shape to the composition. Hydro-electric power is, however, another story.

An aesthetic of water

To the purist, form is function. But the form of function can be diverse yet the function itself always effective, throughout most of the categories of structures and buildings. The purist exalts over the form of the VC 10 or Tristar, Concorde, or of the nuclear-powered submarine. Amongst the static artefacts of this age, only two types are comparable in that their form and structure are dictated entirely by the arduousness of their purpose. These two are the long-span suspension bridge and the thin-arch dam. The bridge reached the apogee of its achievement with the suspension bridge spanning 1,000 metres or more; the dam similarly with the thin concrete arch, impounding a head of water of perhaps 200 metres. Both are unrivalled, except by each other, in scale and daring and in slender beauty of line and proportion. Sir Misha Black has stated that the Severn suspension bridge is the finest artefact of the present age, and is the equal of any work of any living painter or sculptor; I would like to show him the arch dams at Pieve di Cadore or Lumiei in the Carnic Alps, or maybe those in the Massif Central. The arch dam reached its peak with the work of Semenza of Italy and Coyne of France. The 'cupola' arch, a thin, overhung structure, is the most daring as exemplified by Semenza's dam at Lumiei. The ski-jump spillway adds to the dramatic effect, especially if it is placed over the roof of the power station as Coyne chose to do at several dams in the Massif Central where he had little space in a narrow valley. My own choice of excellence is Pieve di Cadore dam (Plate 10). It is not a cupola arch and is less slender than they are, but is unique in another respect which is equally an expression of structural boldness. The radius of the arch is larger than for any other dam; this gives an urbanity of line which makes the dam blend aesthetically into the comparatively wide valley of the Piave, and with the contorted shapes of the peaks of the Dolomites around it. No one sings the praises of the dams in this region, however, because of the disaster at Vaiont reservoir (*see* Chapter 10).

Most of the world's large dams – there are about 10,000 of them – have been built in the last quarter of a century. The massive scale and heroic purpose of the dam can be expressed in concrete by gravity, buttress or multiple arch dams as well as by the arched structures already discussed. The traditional British dam was the earthen embankment – a modest, unassuming structure, perhaps almost unnoticed if it grew grass on its downstream face. This type of structure is still popular and can be very large. In the last ten or fifteen years however the rock-fill embankment has become very popular amongst designers of dams. This movement is, in some respects, a triumph of logistics over intellect. It is now easy and cheap to move mountains of rock. The austere intellectual triumph of the arch dam has been superseded. The rock-fill dam, though undoubtedly impressive and even awe-inspiring at times, is not the equal aesthetically of its predecessor. From an engineering point of view, the difference intellectually is more apparent than real, since a great deal of profound thought and experience has been contributed to devising what is seemingly a heap of rubble. There is considerable subtlety inside the heap! Not an unusual result for a civil engineer, whose finest brainchildren – like difficult foundations – look drab if one ever sees them at all.

Genissiat, the largest French hydro-electric station on the Rhône, and the largest by some margin in Europe when it was completed a few years after the war (but not much greater than a twentieth of the capacity of Krasnoyarsk), is water power's great monument of the 'cathedral of power' ethos. Its concept is unusual in that the dam and power station are united visually into one structure, giving a sculptural effect inspired by functional necessity but carried a stage beyond that by architectural vision, for the power-station roof forms an extension of the dam. The machine hall of the power station is laid out in a slight curve along the downstream face of the dam, and neither inspiration nor expense were spared to give it a lustrous elegance (Plate 12). On the right bank, the jet of the ski-jump spillway thunders as it cascades harmlessly into the river (Plate 11). Pride in power has not been expressed in such a grand manner since that period, although it has had other devotees.

A trend which first became popular in Sweden is to put the power station underground. This was logical and saved money and was not done from fear of war, although maybe that fear had an influence in the first place in directing attention to the method. The underground power station offers its own special problems and attributes to the architect. Areas of wall are left bare, and the uneven surfaces of the native rock contrast with other, smooth and brightly painted surfaces and with the shining precision of the electrical machines and the overhead crane.

In Italy, marble has been used, rather too adventurously sometimes, to decorate the power stations, and there are also murals in some of the stations of the fifties and sixties. Some of them are hideous, but the typical power station of this period has a certain richness and style.

The typical run-of-river scheme has an entirely different visual context. Barrage and power station are often combined, most often in a wide, flat valley dominated by the river, now at a more mature stage than the mountain stream. An attractive lightness of touch has been achieved in some schemes of this kind, especially those where modern sluice gates which do not require overhead bridges and lifting gear have been used.

In some of these schemes a barrage impounds the river and diverts water into a canal, which may be several miles long. The power station is at the other end, or at some point along the canal, and after having been robbed of its energy the water is returned to the river. The aesthetic virtues of a scheme of this kind are not improved if the spit of land between river and canal is used as a tip for the spoil excavated from the canal, as was done for some of the French schemes on the Rhine in Alsace. The grandest scheme of this kind is still today, I think, the Beauharnois power station on the St Lawrence river (Plate 14).

The structures themselves are not the only visual consequence of power generation, as the tree planting at West Burton demonstrates. Architecture of the landscape is inevitable if a responsible attitude is taken to schemes of the scale typical of modern power generation. Reservoirs make perhaps the most significant change. The lakes at Aswan and Kariba are inland seas. The mountain reservoirs of other hydro-electric schemes are almost puddles by comparison, but they are often very attractive and add a new and often welcome interest to the valley and the area. Some reservoirs are unsightly, however, if they are drawn down and have muddy shores.

The naturalists' complaint

Over the last few years there has been a remarkable storm of clamour by 'environmentalists', derogating most of the constructive works of the modern world, by criticizing, sometimes inaccurately, their destructive or harmful side-effects. It is important, I think, that any criticism of this kind should be well-informed and well-directed. Such effects, of course, need to be eliminated, but criticism itself gets a bad name if its quality is poor, and the movement could then work in a direction opposite from the one intended.

At a recent congress where environmental effects were discussed, one of the speakers called the protectors of the environment 'naturalists', the old-fashioned name for them. A naturalist, I think, would be rather critical of this chapter so far. It is rather like disputing with a vegetarian; if he says 'I don't like meat', then that is the end of the argument. To the vegetarian naturalists, I have been presenting a cuisine of architectural steaks and roasts. Even if one dislikes the power station or the dam, each serves an objective purpose and is central to the scheme of living, so some compromise between dislike and necessity must be made.

Pollution and architecture we have already discussed, and the chapter that follows could well be read to appreciate objective purpose. There remain other topics, for example planning and amenity. Which places should be left virgin and untouched? Which others should be the centres of industry, of housing, or sites for power stations? These issues take us rather outside the scope of the book since they touch on the values and purposes of society as a whole, but perhaps the book itself will illuminate some of the arguments when such an issue relates to a power station.

For dams and river utilization, there is another field for discussion. What effects do these river works have on fish, on the biological balance of plants and life in and around the river, and are there harmful effects that were not thought of when they were designed and built? The congress* to which I referred discussed just that subject, with a panel of engineers, a panel of specialists on environmental aspects of rivers, and speakers from the floor of the conference. Development projects are spreading faster than the efforts being made to anticipate their consequences, and the scale of man's interference with nature is unprecedented. One-tenth of the world's stream flow is regulated by reservoirs; two-thirds could be by the end of the century. Few, if any, schemes have been comprehensively examined in the widest ecological terms.

I was impressed, first with the seriousness with which engineers treated these criticisms; that such clearly beneficial works should be attacked, well, well!, but if there was a reason, let's get to the bottom of it and put it right! Second, there was a large measure of agreement between the two panels. When it came to specific problems, the scientific side was regretful that its disciplines were less advanced than those of the engineers; at the stage of research rather than solutions.

All these problems will rest on the engineers' shoulders. The engineer for a dam will have to know when to employ a specialist environmental scientist, and how to build that specialist's recommendations into his

* Question 40. Effects of dams on the environment. Eleventh congress of the International Commission on Large Dams, Madrid, June 1973.

scheme. In the broadest terms, the world has to manage its rivers; the weight given in that management to all the various factors – of profit, of utility, of health, of aesthetics, of the conservation of nature and wild life – will depend on the wisdom of the managers and the pressures put on them by others. The best form of management in some cases is to leave nature alone; in others, to regulate every drop of water. Environmental problems, when they exist, can generally, like engineering ones, be solved, although there is likely to be a 'trade-off' between one aspect and another. Specific cases generally involve a mass of detail before they can be understood; however, one well-known dam is discussed in Chapter 11.

I left that meeting in a rather optimistic frame of mind. There was clearly progress to be made, although one or two problems seemed insoluble. Overall, one can cite schemes that are hamfisted or lopsided in some way, but many more that are not, and to take the view that mostly these works are harmful is not only uncharitable but quite untrue.

Any form of development replaces nature by human artefacts. But in some places the human intrusion is especially harmful. River estuaries are places that tend to suffer particularly. Again, we are back to issues of planning and amenity. There is no reason why balanced compromises should not emerge – clean water, and mud flats specially created for wading birds, say, to balance a purpose such as the reclamation or enlargement of a port.

Energy comes into the picture in diverse ways. Mining has left a tradition of tips and scars – it need not have done; reclamation is feasible. Coal-burning power stations make large quantities of ash (pfa). This material is beginning to become of value in its own right as a material of construction, and it is likely that its disposal will not cause any problems in say five or ten years time. In passing though, I would like to mention one scheme that illustrates the principles that I have been discussing.

At Peterborough there exist large tracts of land laid derelict by quarrying clay for brick-making. To the north-west is the Trent valley with its power stations, all making great tonnages of pfa. Nearby there is also a sugar-beet industry, that produces tons of soil as waste washed off the beet. The derelict land is being reclaimed with pfa, and then top soil added from the beet factories. Special trains run between the power station and the scheme's unloading station, where the pfa is made into a slurry with water, and pumped to the pit being reclaimed. Progress is already becoming rather slow, because so much pfa is being sold.

Lastly, ccology with a difference. People are displaced by reservoirs, and have to be compensated or rehoused. The village of Tignes in the French Alps was a noted case, and after vigorous opposition most of the Tignards accepted compensation. One of them was carefully watched by the French police as he tramped from town to town, sleeping rough. He was a tramp with a difference, and had a fortune in francs in his bag. I never heard whether he spent it, or just kept it as a pillow.

9 Prospect and purpose

Most people, they say, eat to live, but some live to eat. Energy has its analogy. There are some cases where discussion of methods of use, or of pollution, or of economics, is not enough. The most striking case, and it recurs widely throughout the world, is where a river is controlled and exploited. Frequently, the wealth created by utilizing the hydro-electric power of the river is financially the key to the whole operation, but it is not merely a source of wealth and a subsidy that is involved, but a broader philosophy of utilizing natural resources. The purpose of this philosophy is to look, in the round, at all aspects of developing a river catchment, or a region, as the case may be; to investigate how they interact, and to foresee the results of such interaction within a balanced scheme, profitable and useful, yet in harmony with the aspirations of the local people involved and with the land, the river and the ecology of nature around them. It is a philosophy that emphasizes the interdependence of specialists in widely different disciplines; each plays his part within the comprehensive purposes of the scheme.

The archetype is the Tennessee River, and the great programme of works carried through by the Tennessee Valley Authority (T.V.A.) starting in the 1930s. The valley – an area nearly as big as England – was in a dire state of depression, its people afflicted with poverty, unemployment and malaria, its natural balance ruined by soil erosion. The forests had been stripped, and erosion of the unprotected soil of the steep hillsides filled the rivers with silt, so that navigation became difficult, and floods became worse. What was proposed was an overall programme of social and economic management to revitalize the valley; the key was controlling the river and exploiting the wealth of its hydro-electric resources, but although the key, it was still only one task amongst many.

According to Herbert Agar* the success of this venture was 'a genuine discovery in social planning'. This is how he sums it up:

> The Authority was not handed simply a task of engineering, like the Panama Canal or the Boulder Dam. It was told to remake the economic and social life of a vast underprivileged community: through cheap power, land reclamation, reforestation, flood control, diversification of agriculture, terracing of hillsides, encouragement of

* Foreword to Henry Brillings, *The Power and the Valley*, Rupert Hart-Davis 1953.

animal husbandry, cheap transport through restoring the navigability of the river, and abundant vacation-sites on the lakes which would form behind the new dams . . .

We felt the world was against us: the power companies, the railways, the timber interests, the makers of fertilisers, the inert followers of Messrs Coolidge and Hoover who felt that anything the federal government attempted must be wrong. . . . How can one convey the excitement of that fight, which we seemed always on the verge of losing? Dam by dam we battled. . . . The T.V.A. may have sold power at an uneconomic price. When the capital costs of a project are allocated among the many subtle and uncommercial activities listed above, who can determine the exact price of the consequent electric power? But such wrong can be cured with experience, whereas the inhabitants of the Tennessee Valley could never have been cured without this investment of the whole nation's energy, money, and imagination. . . . Perhaps the finest and most hopeful achievement of the Authority is that the citizens of the Valley regard their new society, which has flowered in twenty years, not as something imposed by 'reformers' from far away, but as something which belongs to them, which they helped to create, which in many cases they moulded and shaped according to their local customs and traditions. They were never pushed into accepting an 'improvement' until their objections had been removed by discussion and experiment, and their conservatism overruled by their own experience.

T.V.A.'s battles now rest in the past, and the valley has moved on to a new phase of life, albeit dependent on its dams and river management. The genuine discovery in social planning has been adapted and applied across the world, and indeed others claim it as theirs, but that is less important than its successful application. Unfortunately the discovery has also been forgotten in the building of a few great schemes in recent years, which the founders of T.V.A. would have found unbalanced in some of their aspects, even if they had been overawed by the engineering advances in others.

The wealth created by one large dam, strategically sited at the head of a river catchment, thus creating storage to control the river's flow and a head with which to generate power, can be immense. For example the annual output (i.e. crops and electrical energy) of the great system of power stations and irrigation canals and distributors commanded by Bhakra Dam in India is stated to have a value equal to the capital cost of the scheme. Not surprisingly, multiple-purpose schemes abound on the rivers of the world. Generally power, irrigation, flood control and navigation are the principal purposes. There may not be the depth nor the diversity of social purpose that T.V.A. demonstrated, but there is unquestionably in each case a great natural resource to be managed, which creates great wealth, serves fundamental needs, and functions with a natural rhythm, like cropping the land, if the management is effective and farsighted.

A development of this kind generally excites political controversy, and becomes more or less a part of the political scene at some time during planning or construction, even if only as the brainchild of an ambitious politician. Solid practical utility generally limits controversy to flashes of argument round the edges of achievement. The greatest in size and scope of any of these developments of the world's rivers is the Indus Basin scheme in Pakistan, which is unique politically as well as in its physical size. In this case the scheme itself is a physical expression of a political aim, for it makes workable the separate existence of India and Pakistan. It is based on the Indus Waters Treaty of 1960 which defines how the waters of the five rivers of the Punjab are divided between the two countries. This great catchment is a unity from an engineering point of view, and logically would be embraced by one great scheme of conservation. The treaty splits it into two separate systems, one in each country; waters traditionally used within Pakistan but now diverted in India for irrigation, have been replaced by link canals from rivers farther north, and these link canals in their turn are kept supplied by two great reservoirs, at Mangla on the Jhelum and Tarbela on the Indus. Here lies an interesting comment on the virtue of least-cost engineering solutions to economic needs. In purely physical terms the link canals are unnecessary and the reservoirs could serve more effectively as part of one catchment-wide design. The penalty in cost for dividing into two what ought to be one is probably immense, but has been absorbed within the astronomical sums in the accountancy for the two systems, which is not surprising bearing in mind the claim made for the wealth generated by Bhakra Dam which is the keystone of the Indian system. In the Indus valley, life depends on irrigation, and clearly the division into two was a practical and necessary measure. Politics set the scene in which a life depending directly on engineering, and hence the engineering itself, could flourish; independence was the first step, and perhaps the least objectionable component of its cost was the premium for dividing the irrigation system into two. As one observer put it, 'Water follows immutable laws of nature. Man makes his own laws as he goes along – and immediately breaks them. I prefer water.'*

* A. M. R. Montagu, introducing Aloys A. Michel, *The Indus Rivers: A Study of the Effects of Partition*, Yale University Press, 1967.

Transport and the motor vehicle

Just as the ability to control a great river and extract its energy can be the starting point of a social movement like T.V.A., so the ability to mount a heat engine on four wheels and make a horseless carriage touches deeply rooted human motives. The motor vehicle outclasses the great thinkers, the great revolutions, and the great battles of history in the changes each has wrought. It inspires by making its driver as free as a bird, and flattering his sense of power; it conquers him completely when it can be put to work to make money as well. On the scale of national economies it has had a precisely analogous effect; no other artefact so costly and so desired wears out so quickly, and its manufacture has become an economic base. Its side effects are as shatteringly destructive as its virtues are shiningly constructive. That it fouls the air we have already discussed at length in Chapter 6. Its most destructive effect is that it kills and maims on a scale that would cause torrents of anguish if it occurred in a war.

The high figures of deaths and injuries on the roads seem to be accepted with surprising complacence by society at large, in each and every country. The unnatural mingling of people and vehicles on most roads and ubiquitously in towns, equally is accepted with complacence, even though the noise, the danger and the dirt of the vehicles are destructive of any charm and character that the town has to offer. These things didn't happen the same way with railways, basically because the vehicle and its track were, and are, considered a unity in the world of railways, but are completely divorced from each other in the world of road transport. The distortion so caused is the underlying reason for many of the injurious side effects of a highly successful form of transport.

That same broad philosophy and interdependence of different disciplines which made T.V.A. a success broke down when the motor car came along, and got lost in a traffic jam somewhere. In no country has the use of motor vehicles been related, as a matter of policy, to the roads on which they will run. It is now understood how a large town needs to be reorganized and reconstructed to cater for the motor vehicle; expense, misunderstanding and resistance have prevented much progress being made. The motor vehicle remains the bugbear of town and country planning. The great failure of planning in Britain has been that its needs have been ignored or misunderstood. Subservience of all other needs to it has created the motorized city like Los Angeles, the unbalanced excesses of which are abhorrent to many sensitive souls.

These facts are overlain by the desire of families and individuals to own their own personal vehicle; here lies a social motive of great strength, dependent on lavish use of energy. It is idle to discuss just that aspect, and its derivative, the atmospheric pollution caused by vehicles, if one ignores the deaths or the fundamental inadequacies of planning for vehicles especially in towns. Traffic, like energy, responds to social needs; the character and quality of the age shapes those needs, and so determines the character and quality, in a social rather than a technical sense, of its technology.

Even then, society may be improved in spite of itself, by the habit of using its technological tools and handmaidens. The only city that I have ever been in that smelled of horses rather than petrol and diesel, is Konya in Turkey. In 1967 when I was there, horse-drawn carriages plied for hire in the streets and there was scarcely a motor vehicle to be seen. In the vast agricultural plains around Konya, the horses running with other livestock were sleek, beautiful creatures. In Konya, the cab horses were pathetic bundles of skin and bone, starved, some with sores, some partly lame. It is better to abuse a mechanism than an animal, and the plague of motor vehicles has some side effects to its credit. As for the smell, horse dung in moderation has its advocates, but they are not a powerful lobby.

An excellent summing up in a more academic sense of these general purposes and how they relate to energy usage has been given by Mr A. Silverleaf, director of the (British) Transport and Road Research Laboratory, in his opening paper to an international conference on the future development of airports.* Here is an extract from his introductory remarks:

> Transport is taken for granted as an essential aspect of civilized living. Mechanical transport has given people new freedoms: mobility which allows them to change their patterns of activity, to fashion them on a world scale, to think and behave in a way previously completely unknown. However, it would be facile to assume without question that the changes of the past century are irreversible or irrevocable, or even that the trends since the Second World War – a period which has seen the extraordinary blossoming of air transport – will continue unchanged for more than another decade. Quite apart from environmental arguments, the basic justification for transport has been questioned in ways which may now seem trivial, almost absurd, but which may yet prove to be significant. There are those who argue that the apparently insatiable demand for transport reflects a fundamental imbalance in society: that far from being proud

* Extract from paper presented at the Fourth World Airports Conference. (Institution of Civil Engineers, London 1973) by A. Silverleaf, director of Transport and Road Research Laboratory. Crown copyright.

of the ability to move people and goods with ease and at will, the desire to do so should be regarded as a symptom of failure to organize society efficiently. A well run society minimizes the need for movement and does not use resources and skills heedlessly by mindlessly providing ever more means of transport simply because someone somewhere seems to expect them to be available. On this argument transport restraint rather than unlimited freedom may be a blessing if it forces present patterns of activity to be re-examined and refashioned in ways which make life less restless – and perhaps more satisfying.

Naturally, such radical concepts do not form a realistic basis on which to consider the probable world demands for transport only a decade or so ahead. Today transport is a major activity. In Britain transport now consumes about 20 per cent of the total energy directly available for all purposes, with industry and domestic and ancillary users both consuming about 40 per cent. In the U.S.A. it now absorbs more than 30 per cent with industry taking almost 40 per cent and domestic and other users about 30 per cent. Before the year 2000 the situation in the U.K. may resemble that in the U.S.A. today, and in the U.S.A. out of a much increased total consumption of energy, roughly equal proportions will go to transport, to industry, and to domestic and other users. North America, Western Europe and the U.S.S.R., which together contain rather less than one quarter of the world's population, today consume almost three quarters of the world's energy; probably more than 20 per cent of their share goes to transport.

The main user is of course road transport which dominates almost all aspects of overland transport today. In the U.K., road transport takes almost 80 per cent of the energy input and air transport almost 15 per cent, with rail and other users taking less than 10 per cent. Here it may be worth pointing out that road transport is the most efficient overall user of energy because the direct conversion of a primary fuel such as petroleum to useful work by the internal combustion engine is considerably more efficient than processes involving electricity generation and distribution in which energy losses are substantial. The demands of transport on energy – particularly in the form of liquid fuel, either natural or synthetic – typify the way in which transport is a major user of resources. It is equally a user of materials and of people. Indeed, in most developed countries transport activities as a whole, including those associated with the production of vehicles and the construction of transport infrastructure, account for almost 20 per cent of total national output.

A fresh look at electricity supply

The advantages of large scale in improving the economy of generating sets, together with the advantages of interconnection, are generally accepted and universally practised. It is therefore a surprise to learn that an opposing point of view is held by some engineers. The reasoning which follows is due to Professor Murgatroyd,* who draws attention to. the importance not of advanced engineering at the frontiers of knowledge, but to the staple variety, specifically the large (say 30 MW) diesel engine, one of the most reliable forms of power plant, which surpasses in efficiency all but the most sophisticated turbines. If such sources of energy were used locally to generate electricity for the low-voltage distribution networks already in use, the transformers and high voltage transmission lines could be dispensed with and with that saving (i.e. in cost and in electrical losses), the cost of electricity would be less than from conventional centralized processes, he suggests. A fuel like natural gas adds to the advantage of this approach, which would permit step-by-step growth in phase with demand, and economy in the use of capital – a costly item. Further it would be possible to find a market for the waste heat from some of the diesel engines. There would be difficulties in load-matching, heat-storage and dumping heat, and of noise, pollution and control, but in the professor's view the greatest problems would be legal and administrative ones. Competition between this system and the accepted, centralized one, would help to assess them both, but the monopolistic set-up in Britain makes such competition very unlikely.

If a small prime-mover – perhaps a fuel cell – could be developed, this approach could be scaled down to individual houses, each with its own generating set, served by oil or gas mains which are cheap to install, and using waste heat for central heating and refrigeration in the house.

Total energy

A passage in the professor's address which sets in perspective very neatly the remarks made in Chapter 4 about the combined generation of electricity and heat is best presented in his own words.

> Waste heat has been used for many years in large industrial plants which generate their own electricity. Recently the concept has been greatly extended into the commercial and domestic field, especially in the United States, under the rather glamorous name 'total energy'. The concept of total energy implies taking a single fuel and transforming it into a number of energy-intensive end-products, such as mec-

* 'Forty years of thermal power', by W. Murgatroyd, Professor of Thermal Power. Inaugural lecture 1969, Imperial College, University of London.

hanical work, electricity, heat, refrigeration and drinking water. I have already mentioned the difficulty of doing this on a very large scale; for this reason the use of uranium in total-energy plants appears to be limited to a few major projects such as regional water-purification systems. For the same reason, coal is not used in total-energy systems, although, of course, a success with small fluid-bed boilers could make coal very competitive in certain areas.

There seems to be a growing world-wide market for total-energy plants of varying degrees of sophistication in the range of sizes for which diesel engines, gas engines and gas turbines are the most appropriate power plants, and the increasing availability of oil and especially of natural gas has given an impetus to plants of this kind. Incidentally a large modern ship is a superb example of a total-energy system, having a requirement for mechanical power, electricity, process heat, refrigeration, air-conditioning and even desalination. There are usually difficult problems of space and layout to be overcome.

His conclusion links again with that interdependence of different disciplines which was the key to T.V.A.'s success.

We must learn (he says) to deal more effectively with thermal power systems as a whole, from the fuel supply to the required end products in whatever forms the heat and work are needed . . . there is a danger of over-concentration on components rather than on what is required for the whole system.

In the palmy days of coal, that lesson had been learned in many of the Yorkshire mills, which practised the art of total energy, even if they did not know what it was called. Coal would be burned to yield gas for lighting, steam for mechanical power through a steam engine and heat. Times of depression helped to kill the practice, if for example one floor was let and the electricity company offered an attractive tariff to the tenant.

There is a scheme in England of a similar nature in service in Nottingham. The turbines in an old, small industrial power station have been refurbished, and its boilers serve to supplement the heat recovered from a plant where the city's refuse is incinerated. The electrical output works the incinerator plant and the heat station. Eventually a sizeable area of the city will be supplied with heat, which is the product that is sold.

The gas turbine is an advantageous machine to use as the basis of a scheme of this kind, which could easily be applied in a small heat-power station to serve a large-scale building redevelopment, of the kind one sees so frequently, thus replacing numerous boilers sited in individual buildings and operating relatively expensively and inefficiently.

Alternatively, and on a scale similar to present-day electricity generation, the intermediate-take-off condensing turbine offers advantages. Steam is bled off at an intermediate point, used to heat district heating mains and then returned to the boiler. The turbine thus operates conventionally at the low-pressure end. If the district heating system has a balancing heat store, the bleed of steam can be stopped when the electrical peak load occurs, so increasing the power generated, or it can be stopped entirely for the summer season. The cost of this heat is so advantageous that the cost of the district heating mains can be easily accommodated within a favourable heat supply tariff. A further suggestion (from Mr R. M. E. Diamant in a paper to the Institution of Heating and Ventilating Engineers) is that where a power station is remote from a city a single-pipe main, carrying superheated water at 205 deg.C, would carry heat economically and advantageously to the centre of consumption. Yet these lines of development remain neglected, even though reduction of heat losses is more than justified economically.

Still another approach which lacks entirely the glamour of advanced technology, but rather is based on prudent housekeeping, is to take the older, less efficient, generating sets in the electricity supply system and convert them to back-pressure working. Heat would then be recovered from the exhaust steam for district heating, which would be considered the primary product, and electricity the byproduct.*

Energy losses

Domestic heating and air conditioning make major demands on fuels to the extent that their 'incessant growth is almost a dissipation' in the words of one eminent critic, who pointed out the vast improvements possible since standards domestically are so low. The reply was as follows:

> We are all seeking in our homes not only seclusion, security and protection against the outside and the weather, but also we are looking for contact with the outside, with the heavens, with nature and the open air. It is a reaction which is a consequence of the extensive separation of a large part of mankind from its natural surroundings. It would be a fantasy to believe that a basic tendency of this nature could be affected or even checked by considerations of the conservation of energy.

* Norman Jenkins, 'Cheaper heat and power', *The Consulting Engineer*, April 1973.

These exchanges came from the conference on energy losses held by the World Power Conference in Lausanne in 1964; that is my source for stating that 80 per cent of the world's production of energy is losses. The main conclusion one could draw from the conference was that no one was interested simply in reducing losses – that was incidental to other aims. For example, British purveyors of domestic heat – by oil, gas or electricity – were each anxious to show that they offered the cheapest method, and greater efficiency was a ploy in that argument. Once they've got you hooked, the more you buy from them and the better they like it.

The slow rate of improvement in reducing heat losses was singled out as a characteristic of the energy scene, but financial and economic reasons limited the reductions, not technical ones, it was stated. Conclusion: the extent of losses is gigantic and the limits of their reduction very restricted. So the conference's reporter summed it up, and took comfort – indeed fashions do change in a decade – by contrasting man's efforts with the unlimited waste of energy in all the natural phenomena around us.

10 Accidents and disasters

The builders of medieval cathedrals were always striving for perfection by stretching out the limits of height and lightness of their structures. The occasional collapse, even if disastrous, became also a source of experience on which to build again successfully. In spite of the wealth of knowledge today, in spite of advanced methods of mathematical analysis, there is not so much difference. Experience is the great teacher. Its bitterest lessons are sometimes the most valuable, and there is still this element of learning from failures. Failures of this kind are not necessarily caused by mistakes, nor by simple carelessness, nor by omission in applying the designer's skill or the constructor's care and attention. New circumstances may be met, which perhaps cannot be predicted in all their complexity. Failure may be tragic, but not culpable. In other cases, carelessness may be ruinous. Just one or two examples are selected from the world of energy.

There is no structure built by man for which its designers carry such heavy responsibility as the large dam. It can hold back a sea of water, impounded to a height of up to around 300 metres – a gigantic store of energy of enormous destructive power. Gruner, the Swiss engineer, concluded a paper on dam disasters by writing:

> Every dam which impounds water presents a potential danger which should neither be under- nor over-estimated. The risk of sudden disaster is for ever inescapable, and while knowledge and vigilance may reduce such a risk it can never be entirely banished. All man-made works carry within them the seeds of their own decay and eventual destruction.

Two failures of dams in recent years have especially drawn attention to these dangers, by the scale of the catastrophe they inflicted and by the effects of the failures on engineering thought. They both relate to arch dams, and respectively to the arch dam's two greatest exponents, Coyne of France and Semenza of Italy. In the 1950s many arch dams had been built; not one had failed, and the arch dam was considered to be the safest and most reliable of all types of dam, and the one most favoured if the site permitted its use. But in 1959, Malpasset Dam in the south of France, a relatively small dam designed by Coyne's firm, collapsed suddenly when the reservoir first filled. The flood which resulted swept destructively through the town of Fréjus, killing many people and causing damage estimated at £22 million.

The arch dam exerts a side thrust on the rocks at the sides of the valley where it is sited, just as the arch of a bridge thrusts sideways on its springings. If this thrust cannot be resisted by the supports without deformation, then the stability of the arch is lost. Thus it is fundamental that an arch dam requires a valley with sound rock on both banks that will provide a solid, immovable springing against which the dam can thrust. Investigation after the failure at Malpasset revealed a complicated and improbable set of conditions in the rock. One effect was that the load from the dam remained concentrated, and did not spread out, to some depth below it. Planes of weakness – from various causes – defined a wedge-shaped region of rocks. When the reservoir filled, the arch thrust, the wedge lifted, and failure was sudden and catastrophic.

The tragedy at Fréjus was to be repeated on a larger scale and with a ferocity that surpasses comment, when the village of Longarone in the Carnic Alps was obliterated one night in October 1963; about 2,000 lives were lost. Semenza's Vaiont Dam was the highest dam in the world at that time, well known and a triumph in the van of engineering progress. It was built in a valley of incredible steepness, and was a work of enormous daring and accomplishment. Landslips down the steep slopes beside the reservoir were expected from time to time, and instability of the slopes was expected and monitored. At the time of the disaster, engineers were on the mountainside and had observed that it was moving, but it was considered that there was no cause for alarm.

That conclusion was tragically incorrect. The side of a mountain – Mount Toc – slid, at a speed probably around 90 km/hr, into the reservoir. The effect was like dropping a brick into a bucketful of water that is only slightly bigger than the brick – more than half the water plops over the edges of the bucket. Vaiont Dam is almost 270 metres in height; it had 120 metres of water over its crest that night. The violence of the flood in the steep valley defies any sort of description. The dam was on a tributary, and Longarone was beside the larger river, upstream of the confluence of the two. The flood waters swirled up the main valley and then retreated, leaving an expanse of scoured mud where the village had been.

Even after that, it is a matter of importance to engineers that Vaiont Dam withstood the tremendous overload of the surge of water over its crest, and still stands, useless but unharmed – a bitter confirmation of the high structural reputation of the arch dam. It would have made little difference to the tragedy if the dam had failed; the reservoir is completely blocked by the fall, quite close to the dam.

Adverse geological conditions, weakening of the rock by hydrostatic pressure from the deep reservoir, and further weakening by ground water recharge (there had been heavy rain) are given as causes. Water penetrated deeply into rocks that had been well drained before, and at high pressures – around 250 tonnes to the square metre maximum. A phenomenon well known in soil mechanics is called liquefaction. Under certain conditions of imposed load, loose, fine-grained sands or silts behave just as if they were liquid, until the load is relieved. The tragedy at Aberfan in South Wales, when a colliery tip ran down the hill and engulfed a school, was an example of it. It is thought that some similar explanation is necessary to account for the great slip of Mount Toc.

The volume of rock which fell was estimated to be about a quarter of a cubic kilometre, and it extended over a length of about two kilometres, and over a surface area of two square kilometres. This mass of rock is ten times larger than any other landslide in historical times. The engineers' trial holes on the mountainside, although deep by any normal standards, had not penetrated nearly deeply enough to discover the slip surface on this gigantic and unbelievable scale. According to one authority, the entire landslide was over in fifteen to thirty seconds; at that time the force exerted on the dam was four million tonnes.* A clean surface of fracture was exposed behind the slip, sloping at 60 deg and 1·8 km in extent. The trees and vegetation on the fallen top surface were relatively undisturbed.

These two disasters led to profound changes in world engineering opinion. Terzaghi, the father figure of soil mechanics, said that it would be impossible to carry out site investigations to the degree where such conditions as caused the Malpasset failure could be detected with certainty. Others were more confident, and while they increased very considerably the investigation of abutment rocks, continued to build arch dams. Another result was a sudden burgeoning of interest in the science of rock mechanics. Much valuable knowledge applicable to tunnels and foundations as well as to dams has resulted, but I have never seen any published result that could be applied to the stability of reservoir banks of topography and geology similar to those of the Vaiont reservoir. A site so steep will never be chosen again for a reservoir of such great depth.

Spilt milk

The safety record of nuclear power has so far been very good. The only accident on record that had potential danger for the public at large (I believe there have been accidental deaths amongst reactor staffs in the

* G. A. Kiersch, *Civil Engineering* (*USA*), March 1964.

U.S.A.) occurred at Windscale in Cumberland in 1957. Windscale and Calder Hall are adjoining, but Windscale is an experimental and military establishment, and the reactor which caused the trouble was not a power-generating installation but was used for the production of plutonium for the military.

Not to put too fine a point on it, the reactor caught fire. It was cooled by air; the air was not recirculated as the CO_2 is at Calder Hall, but was discharged through filters at the top of a tall stack. Radioactivity in the cooling air increased as the fire burned, and some of it escaped in the plume from the stack.

It proved to be a radioactive isotope of iodine that escaped, fortunately one that decays fairly quickly. It was deposited on the fields around Windscale, where dairy herds graze. The cows ate the grass, and the radioactivity which the grass had picked up passed into their milk. It was estimated that the thyroid gland of a child could possibly have been affected by the concentrations of radioactivity in the milk. Milk deliveries from twelve farmers within a two-mile radius of Windscale were immediately stopped, and two days later had been stopped over a strip of land thirty miles long and six to ten miles wide. The news media were full of pictures of milk being poured down drains. It was about six weeks before normality returned to the dairies.

The accident happened when Wigner energy was being released from the graphite moderator of the pile. The graphite builds up energy in its molecular structure, which had to be released by heating it, in this particular reactor, but the heat was applied too rapidly and some of the fuel cartridges failed and started to burn. The trouble started probably on the Tuesday, the fire was affecting 150 fuel-element channels on the Thursday, the reactor was flooded with water and the shutdown fans closed on Friday, and by Saturday afternoon the reactor was cold. The accident had been dealt with promptly and efficiently and no one was hurt. There is a deal of technicality in understanding all the things that happened, but it is also very clear that a magnox reactor would not behave in a similar way; this particular accident does not seem to have relevance in a technical sense to today's or tomorrow's reactors.

The reactor was subsequently sealed off, and another similar one nearby has also been sealed – the first of the radioactive monuments that I referred to earlier. Apparently they are no great embarrassment, and work demanding freedom from contamination goes on in the building which abuts the sealed reactor. The nuclear specialists in fact can see no great problem ahead with their discarded reactors, and consider that they have solved much more acute problems than that in developing nuclear power. The sealed reactor vessel would not be very much of a hindrance on a site still used for power generation, they consider, and on a site reverting to other uses could be covered with earth and virtually forgotten.

The New York blackout

On 9 November 1965 the state of New York, six adjoining states and two Canadian provinces, were without electrical power for ten hours or more; 25 million people were affected over an area of 80,000 square miles. Trains, lifts, public services, hospitals, etc. all came to a standstill. There were few casualties and there was little or no panic. The failure occurred at the evening peak hour (5.15 p.m.) and was caused initially by a faulty relay at a Canadian power station at Niagara Falls. About 300 MW of power flowing into New York along a 345 kV line was thus interrupted; other lines should have made up the deficit but the transfer of load caused them to surge and to trip out of circuit. Thus a cascade of failures was initiated, as power generating systems in the area were subjected to instability and so tripped out as their automatic protective systems came into play. Some of the supplies were reconnected in a few minutes, but it was fourteen hours before all the faults had been rectified. The breakdown, as the *Electrical Times* noted at the time, was a reminder of 'the often-forgotten complexity of today's power systems and of the almost unlimited numbers of possible combinations of fault conditions with which system designers ought to concern themselves. The event was unlikely . . . but never impossible it seems.' It was also a reminder of the great dependence of the modern city on electricity. Several other similar events are on record, but this one was the most spectacular.

Freezing fog and gales

When rain freezes as it comes into contact with something solid, or when fog freezes, the electricity grid is in trouble. Insulators become ineffective; ice can build up on transmission towers, maybe in sufficient quantity to overload the tower to the point of collapse.

Gales also cause trouble, especially if they follow or coincide with icing up. Conductor wires may 'gallop'; towers can collapse. Chimneys seem to have a good record, although some have 'galloped', a habit which is prevented by fitting spiral fins round the tops of them. In aerodynamic terms, the fins control vortex shedding and so prevent resonant vibration. But for sheer spectacular failure the cooling towers at Ferrybridge power station in Yorkshire, one of Britain's 2,000 MW stations, take the prize. According to reports at the time (November 1965), the towers were 'fairly rippling in the gale' and emitted a high whine. They clearly attracted attention, and the moment of failure was recorded photographically (Plate 16).

The failure has an interesting engineering background over which one can dally without remorse, since no lives were lost. The towers (they have been rebuilt) form a group of eight, closely spaced on a pillar of ground unaffected by mining. Because they were so close, turbulence from the four windward ones buffeted the four leeward ones, so increasing the wind loading on them.

Towers of this size had not been built before; they are 375 ft high and 300 ft in diameter. The doubly curved shape indicates, however, that the wall of the tower is a thin shell, and in this case it was merely five inches thick. In proportion the tower was three times thinner than an egg, yet aerodynamically bluff. This failure can be explained in simple terms. The deadweight of the tower causes compression in the shell; the wind load is an uplift and causes tension. The wind loading specified in design was too small, partly because a report had been misinterpreted and no one had spotted the error. When tension exceeded compression the only thing holding the tower together was the steel reinforcement in the concrete, and it was inadequate. These explanations always seem very simple after the investigation, and when the best brains have been deployed to ferret out the answer. This case underlines the fact that design is not just a matter of applying rules. If thought had been given to the way the tower was going to behave when it failed, a simple sum equating the tensile load in the reinforcement with the uplift from the wind would have exposed the inadequacy; but the design of concrete shells is highly complex and there are not many opportunities for such simple checks. Ingenious work was done on the other, similar, towers in the programme to strengthen them.

The Santa Barbara blow-out

Drilling for oil is a hazardous, and indeed spectacular business, a fact which must be obvious even to those who have not seen John Wayne extinguishing a burning gusher with his bare hands. One example is worth while.

At Santa Barbara in California in January 1969, a well was being sunk offshore. The drill reached oil and gas under very high pressure, so high that all the safety devices failed. The well was capped, but the gas and oil, now at full pressure below the cap, forced open cracks in the seabed and escaped. Eleven days later the vents had been sealed after tremendous efforts. Oil had escaped at a rate of about 500 barrels per day (a barrel is 35 Imperial gallons, nearly 160 litres) most of which polluted the sea and the beaches.

Three months later there was little evidence that pollution had occurred. It was pointed out that there were two submarine oil springs off the California coast, one of which flowed at twenty barrels per day, which had been discovered by Spanish explorers four centuries ago. A more striking conclusion was that it was considered better to relieve the pressure, at risk of some pollution, than to leave nature alone. The area is seismically active, and it was considered that an earth tremor could have had far worse consequences than the failure of the well.

Torrey Canyon

The most sensational and widely reported case of oil pollution of the sea must surely be that of the *Torrey Canyon*, a Liberian tanker which ran aground off Land's End in March 1967, with a cargo of 119,000 tonnes of crude oil from Kuwait. In ten days about 50,000 tonnes of oil leaked out from the stricken ship, and then a gale blew up which broke her back, and another 50,000 tonnes escaped. Salvage was abandoned, and the ship was bombed by R.A.F. aircraft in an attempt to set light to the remaining 20,000 tonnes of oil.

This great mass of oil, at least ten and maybe thirty times the quantity spilled at Santa Barbara, was attacked at sea by twenty-four British vessels, off Brittany by 600 French craft, and by thousands of people on the beaches of both coasts. One of the principal methods was to spray the oil with detergent to emulsify it.

Later, the British Government's committee of scientists reported that 'the summer brought few complaints . . . there is hardly any memory of the public anxiety which prevailed during the period of crisis'. The loss of life amongst sea birds, of which a few species are especially vulnerable, had been high, and pathetic to see, but the public soon forgot, and any destruction of marine life was hidden from view and difficult to assess. An interesting point is that detergents are toxic to biochemical processes, and the view is strongly held that the detergent – 11 million litres of it used by the British – caused more damage than the oil.

11 Engineering cadenzas

The South of Scotland Electricity Board's power station at Longannet, on the north bank of the Firth of Forth about half way between Edinburgh and Glasgow, is at present the largest power station in Britain (Plates 23–7). It demonstrated its reliability by generating more than half the Board's output during its first winter (1972–3) and it is in all respects a fine example of the state of the art today, built on an ideal site for a power station.

It has four 600 MW generating sets, so its capacity is comfortably more than stations of the 2,000 MW (4 × 500 MW) family south of the border. Soon it will be overtaken in size; the new generation of British power stations will have sets of 660 MW (the first one synchronized in January 1973), the largest with five in one station. Larger sets still – in the range 1,100/1,200 MW – are now being built, some by British firms for installation in the U.S.A. and one in Russia. To recall, less than twenty years ago there were few sets over 75 MW in capacity in service in Britain.

The sets at Longannet are cross-compounded and each one has two shafts; there is an identical alternator on each shaft and the two work as one electrically. Steam passes through the high-pressure turbine on one of the shafts, and then is returned to the boiler for reheating, before working through the intermediate-pressure turbine on the other shaft. Then the steam divides, and works through one of four low-pressure turbines, two on each shaft. The set operates at 3,000 r.p.m. For comparison, note that the British 500 and 660 MW are tandem sets – that is, they have only one shaft – and operate with similar speed and steam conditions. The big U.S. machines will take wet steam from water-cooled nuclear reactors, so giving turbine designers a different set of problems; they will operate at lower speed.

Steam is formed in the drum at the top of the boiler, from water already heated in the cooler parts of the boiler. The steam is then superheated in three stages in the hotter regions. The boiler consists of a box-like space into which pulverized coal – coal ground until it is rather coarser than face powder – is blown and burned. Water walls – tubes containing water – and assemblies of tubes carry water and steam through the great box to pick up the heat of the fire. The coldest water is preheated with the lowest grade of heat available, and the last stage of superheat occurs at the hottest region, thus exemplifying the complex circulation of the boiler which pushes such measures to get the highest possible thermal efficiency.

The coal mills – eight to a boiler – are massive and powerful machines. Feed water has to be pumped into the boiler against high pressure, and the design of the feed-water pumps is elaborate and specialized. Air for combustion also has to be handled massively and somewhat precisely. Although it does not refer to Longannet, it is worth returning again to Professor Murgatroyd's address (*see* p. 102) where he refers to boilers:

> The furnace and steam generator is a large box 200 feet high, approximately the size of Canterbury Cathedral, containing several miles of high-pressure tubing. Several complex fluid-flow and heat-transfer phenomena occur in this box: combustion, followed by convection and radiant heat transfer from the hot gases to the tubes, as well as heat transfer from the tubes to the boiling water . . . the chairman of the generating board has said of these boilers, 'The modern large boiler is a clumsy and complex animal and we still do not know anything like enough about the simple problem of burning coal, but within the limits of our existing techniques we seem to be able to build the brute as big as we want to – not that either design or construction is easy, still less that the result is elegant.' It is not surprising that boiler design is still very much an empirical matter and even less surprising that an undergraduate learns very little about it; there are more elegant ways of illustrating to students the fundamentals of fluid flow and thermodynamics.

Coal. The station has an extensive coal store. About half of its coal comes from the local colliery, there being a shaft from it beside the coal store. This shaft is served by an underground conveyor 9 kilometres in length. The remaining coal comes by rail in special trains. The coal store has a capacity of one million tonnes (one-fifth of a year's supply) and can handle on its conveyer belts 24,000 tonnes daily, just about 10 per cent more than the station will burn if it is on full load for twenty-four hours. The coal is graded by big scrapers into smooth, hump-shaped heaps so that it is compact, and air and water do not collect in the heap – a precaution necessary to prevent spontaneous combustion.

Ash. The station is built on land reclaimed from the Forth with ash from the neighbouring power station, Kincardine. Some of its own ash is sold, and the rest is being used for further reclamation. Up to about 20 per cent of the weight of coal burned is recovered as ash, mostly as pfa from the electrostatic precipitators, but about a fifth or sixth is furnace bottom ash.

Feed water. Canterbury Cathedral – the boiler – is a place of remarkable purity. One tenth of one part per million of solid materials in the feed water accumulates into over a tonne of solids in the boiler in a year. The remarkable purity of boiler-feed water is an achievement on which thermal power generation depends heavily, but the fact is not widely appreciated. Water that is completely pure has a conductivity of 0·037 micromhos/cm. The feed water at Longannet is kept below about 0·5 micromhos/cm, and solids below 0·5 parts per million. That standard is not unduly arduous to achieve, but it has to be maintained consistently for large quantities of water, and to do this water treatment plant that is fully automatic and rather subtle equipment for analysis and measurement, which records automatically, are installed. Rather more onerous standards for feed water are specified for other plants where the boilers are more sensitive.

Condensers. Steam exhausted from the low-pressure turbines is condensed in large box-like devices beneath the turbines. They are boilers in reverse, operating at very low pressure, extracting heat from the steam so that it condenses and can be returned as feed water to the boiler. Cooling water – a separate flow entirely from the feed water, it should be noted, and as regards purity, as profane as the feed water is sacred – is circulated in very large quantities and takes away to waste more than half the heat obtained by burning the coal.

Cooling water. The cooling water itself has to be either cooled and recirculated, or else renewed. At Longannet it is renewed. Cooling towers have to be built at sites less well endowed with water, but there is plenty of deep water in the Forth. An intake draws it in, and it is discharged into a canal which rejoins the Forth a mile farther upstream. This configuration was studied with a hydraulic model to assess the effects of the cooling water on the river. The maximum rise in temperature of any considerable body of water, even if the power station runs at full load day and night for several weeks, is 2·2 deg.C.*

Output. The station's output is generated at 17 kV. Transformers then step up the voltage to 275 kV, the transmission voltage. The high-voltage switchgear is in a building next to the turbine hall, from which two transmission lines lead off to serve the south of Scotland.

Control. Elaborate measures have been taken to control the generating sets and the dispatch of power to the grid. The latter operation is under instruction from the grid control centre, but the control room at

* Frazer, Barr and Smith, Proceedings of the Institution of Civil Engineers, January 1968. Paper 7039.

Longannet is comprehensively equipped to monitor and record every aspect of operating the generating plant, the auxiliaries and the coal store.

Aswan High Dam

The High Dam simply is life to Egypt. It has been the centre of engineering controversy, of world political crisis, of intense national pride, and latterly of much unjustly destructive comment around the world.

Traditionally, Egypt depended on basin irrigation. The Nile's flood waters spread over the fields, turning them into ponds; the water receded and the crop sprouted and grew. Perhaps it was the most ancient system of agriculture in the world. In 1902 the first Aswan Dam was completed. 'Though the element we were dealing with was water, we literally set the country ablaze,' wrote Sir William Willcocks, the engineer who designed and built it. 'We made some mistakes and we secured many triumphs; but we went on conquering and to conquer.' The dam stored part of the flood, and released it later for irrigation when the Nile was low, so permitting the practice of perennial irrigation. The old dam has been heightened twice and a power station added to it; it has been a highly successful source of wealth and of crops.

The High Dam does the same thing but on the scale of the whole flow of the river, and with a large output of electrical power. A simplified understanding of the function of the two dams can be gained by examining the figures in Table 6. For the Nile, one measures its flow in cubic kilometres; in an average year (averaged over 92 years of record) it brings 91 cubic kilometres of water through the Sudan and Egypt. The flood starts in July, reaches a peak by August or September, and tails off by the end of the year. The old dam could hold back 5 cubic kilometres for perennial irrigation; the High Dam can store 164, for various purposes.

The first point is that the Nile is uneven year by year, in the size of its flood, and so in its total flow. Sequences of dry years are on record. They brought small crops, drought and famine to Egypt. The High Dam will store roughly an average year's flow (i.e. the live storage in the table) to tide over the low years, and virtually guarantees the mean annual flow; calculating the secured draft is an erudite business, but if not certain in its accuracy is more systematic now than when it was divined by the priests of ancient times. Storage is thus a principal purpose of the High Dam, permitting the replacement of the extravagant basin system of irrigation by perennial irrigation, and vast extension, with confidence, of the land under perennial irrigation. The figures in the table for irrigation are based on a rather conservative

TABLE 6: THE NILE AND THE HIGH DAM

Hydrology and storage	cu km
Mean annual flow over 92 years of record	91
Highest ,, ,, ,, ,, ,, ,, ,,	134
Lowest ,, ,, ,, ,, ,, ,, ,,	42
Mean annual flow, 1869/70–98/99	103
,, ,, ,, 1900/01–60/61	85
Storage capacity of old dam	5
Storage capacity of High Dam	164
of which: silt deposition	30
flood control	44
live storage	90
'secured draft' calculated from	
live storage: for 99% years	91
for 97% years	93
Nile Waters Agreement (1959)	
Prior water rights: Sudan	4
Egypt	48
Reservoir losses	10
Extra flow available: Sudan	14·5
Egypt	7·5
Total	84·0

Note: the agreement was based on the low average flow of 85.

The High Dam	
Length	3,600 m
Height above river bed	111 m
Depth of grout curtain below river bed	180 m
Volume of fill in the dam	41 million cu m
Power station: installed capacity	2,100 MW
number of generating sets	12
annual energy generated	10 TWh
Irrigation (in Egypt only): conversion from basin to perennial	3,000 sq km
additional perennial	5,500 sq km

figure for secured draft, but the additional perennial figure is an increase of about a quarter for Egypt. In addition there is substantial increase made possible in the Sudan, for which no figure is given in the table.

The hydro-electric power is an additional source of wealth. The power station has a capacity a little less than that of Longannet, and with an even flow from the reservoir it could operate at very high load factor. Its estimated output is thus equivalent to burning about three million tonnes of coal in a year. Demand in Egypt has not yet built up to that size, but the reservoir is not full yet. Apparently it will be in a couple of years or so, depending on how generous the Nile proves to be. In addition, the power generated at the old dam, which used to be lost during the flood, has become firm.

What is there against the High Dam? First that it traps the Nile silt which used to debouch over Egypt's fields in the brown waters of the floods. Certainly it does, but close analysis shows that the silt is of rather minor importance to agriculture in Egypt, and that maize, for example, does better without it. The cost of making fertilizer abundantly greater in useful quantity is less than the cost of dredging silt from the irrigation canals. About nine-tenths of the silt actually went straight out to sea with the excess flood waters of the uncontrolled river. The space reserved for storing silt in the reservoir will suffice for about five hundred years. The silt-free waters below the dam pick up a fresh load of silt, and the erosive damage so caused is on the debit side. This effect was expected and has been closely studied. It occurred mostly in the first few years after closure of the Nile, and its rate is now much reduced; a drop in water level of 40 to 80 centimetres has been observed at barrages downstream of Aswan, and some remedial works will be carried out.

Then there are the sardines, which moved away from the mouth of the Nile, and even had the gall to finish up off Israel according to one report. A minor loss of fishing has occurred, but the Egyptian fishery researchers expect to make it up at favourable places along their coastline. The balance sheet of the dam, however, is again massively positive; 18,000 tonnes of fish are being caught in a year in Lake Nasser, and a catch of 80,000 to 100,000 tonnes is predicted in the future.

Bilharzia has been endemic in Egypt since ancient times. Until methods of eradicating it are found it will persist where there is perennial irrigation, and that there must be. Measures can be taken to discourage the snail hosts of the disease, which are being introduced in the new irrigation works. The cost of controlling the disease is estimated by the Egyptians to be about a pound a year per irrigated acre, for molluscide to kill the snails. One province of Egypt is now free from the disease.

Lake Nasser, the reservoir, is in good heart, without malarial mosquitoes and not unduly burdened with weed. There has been a spot of coast erosion at the mouth of the delta with the new régime; it is insignificant. That, I think, covers briefly the main points that I have seen discussed by the naturalists. I have avoided quoting many figures and detailed verification, but it is all recorded in papers at the conference referred to on page 93. There are two engineering points that have been knocked about a bit by the specialists – they are seepage and evaporation losses from the reservoir – and I think it will not be possible to have a definitive answer until the reservoir has been filled for a few years. Performance up to now, however, contradicts at least the direst prediction, which was that the Nile would run dry for a year or two when the reservoir started to fill. Needless to say, the losses envisaged in the Nile Waters Agreement – 10 cubic kilometres a year – are severe enough and dwarf completely all the naturalists' drawbacks in scale. On the other hand, the philosophy behind the dam was vindicated in 1972 when the Nile yielded only about 40 cubic kilometres, its lowest flow since 1913; without the dam there would have been catastrophe.

How much shall I enthuse on the engineering technicalities of the dam, and over a certain byplay between Russian and Western methods (British consultants designed it originally) of building it? Just a few facts. The dam stands on a permeable river bed, and a deep curtain of alluvial grouting had to be formed to make a watertight cutoff beneath the dam. This central vertical watertight barrier continues upwards through the body of the dam as an impermeable core of clay. The cutoff is supplemented by a blanket – literally that – of clay on the upstream side, with a funny shape that can accommodate settlement of the dam should it occur. There are also drainage wells on the downstream side, and in dire catastrophe, if one or even two of these safeguards failed, the dam would behave tolerably; that is a higher standard of safety than most dams can boast.

Another unusual point is that the dam was built in the wet, by dumping rock into the water of the reservoir of the old dam. Special techniques were used to consolidate the sand on which the rock was dumped, and for other purposes. They were novel and proved successful. But the great scale of the work and the enormous volume of fill which had to be placed in the dam generally attracted most of the attention.

The climax of construction was the closure, when the Nile was diverted from its course through the tunnels of the scheme. It was a time of great excitement; there is a film about it which I heard described aptly as 'pure Cecil B. de Mille'. The diversion had to be done before the flood came, or a year would be lost. Figures can sometimes be so revealing; look at Figure 13, overleaf; it shows the progress made in lining the tunnels with concrete, and stops at the end of April 1964, in time for diversion in May. Progress reached an astronomical peak in the last month. Subsequently, the work became slightly less frenetic, but never lost the character of a national proving-ground.

The Forties platforms

Oil has been found beneath the North Sea where the depths of water are around 120 metres or so. It was decided to build towers that would stand on the sea bed in this great depth of water, each with a platform on top of it, above the reach of the highest abnormal wave, and to install and operate the drilling and pumping plant and the pipeline

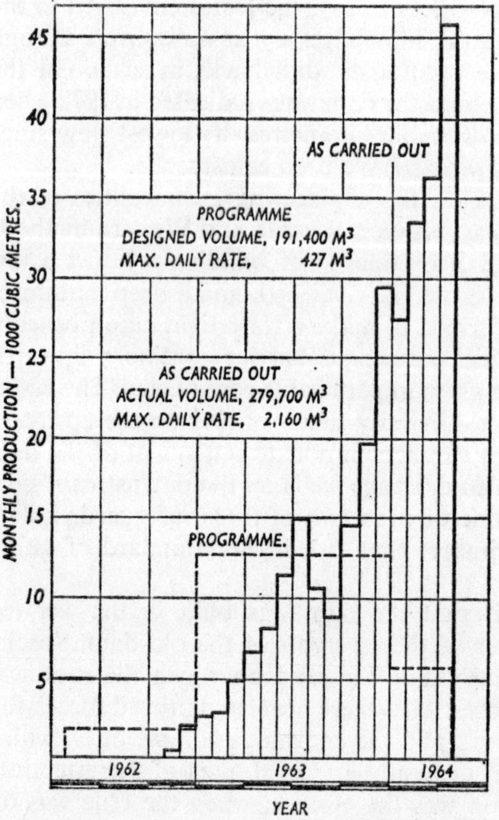

Fig. 13 Progress in concreting the linings of the tunnels during construction of the power station at Aswan High Dam.

terminals from these platforms. Nothing of this size, nor in such arduous conditions of open sea, had ever been attempted before.

Two platforms were being built in 1973 – each on its side on a raft in a tailor-made dry dock – with the intention of floating them out to sea but they are both running a bit late, and this book will be published before the acid test takes place. I can still describe the construction in progress of one of them. It is simply a square tower of steel tubes, pinned to the sea bed by piles, and it is of great size which is its outstanding characteristic. The piles will be driven up to nearly 100 metres into the sea bed with a hammer weighing 250 tonnes, and 18 metres long. Construction has been complex and arduous – it could not have been attempted ten years earlier – and special attention has been paid to upending the platform and the raft on which it will float, by controlled flooding when they get to the site.

The whole venture is titanic in scale and bold in execution so forgive me for just a sentence of armchair carping. The structural concept of these towers seems to me to be utterly crude; I think those that follow in later stages of exploitation of North Sea oil will be radically different. As a starting point, there is little interest in anything except tapping, as quickly as possible, the source of wealth that lies in the Forties Field.

12 Fire, air, water and earth

This chapter, like the last, is a sample. It tastes the great variety of invention which seems to outpace even the relentless exponential growth of people and energy usage. It takes for discussion examples of energy from the sun, the wind, the tides and the earth itself, but first explains a promising possibility for the immediate future.

Further improvements in the economy and the thermal efficiency of electricity generation will unroll as the years pass. Maybe the elegance of a direct method will always be elusive; if there is some latterday Faraday who could make water magnetic and introduce the direct hydroelectric method he would be in as fortunate a position as the man who could find a use for oceans of tepid water.

The immediate possibilities are numerous, but often complicated to explain – such things as combined cycles for gas and steam turbines for example. Basic changes in the system – a total energy approach or isolated generation – are not outlandishly farfetched. On the other hand if the present interconnected system of power generation flourishes some opinions have it that fuel for transport will eventually become hydrogen or methanol. For example, hydrogen would be made by electrolysis at the power station, distributed by pipe as the perfect gas, and liquefied for use as a benign fuel for transport. Improved batteries or fuel cells are other possible approaches, or equally the return of the steam road vehicle, which dropped out of the evolutionary race not from technical inferiority but because of pressures from established industrial interests, and discriminatory taxation.

Combustion on a fluidized bed

Immediate promise is generally considered to be offered by the fluidized bed technique for burning coal. Here is a prospect for superseding those great Canterbury Cathedrals of boilers, without going to the other extreme of the fireball of the fast reactor. If gas is blown through a bed of coal, the coal lifts and behaves like a fluid – somewhat like the liquefaction phenomenon in soils discussed in Chapter 10. If the gas is air, the coal can be burned at about 800 deg.C, a much lower and more convenient temperature than the 1,400 degrees of conventional boilers, and if water tubes are immersed in the bed, much better heat transfer is obtained than with conventional boilers. If the bed is pressurized to several atmospheres, better results still are obtained, the higher pressure improving the heat transfer properties. About half the sulphur in the coal is fixed in the ash, and virtually all of it can be fixed if limestone is added to the coal; oxides of nitrogen do not form because of the low combustion temperatures. Thus the range of coals which can be burned in this way is increased, if atmospheric pollution is a consideration, and there is a further advantage that coals with ash contents greater than the present upper limit of 16 to 20 per cent could be successfully burned in the fluidized bed. For a power station boiler, Canterbury Cathedral would reduce to the parish church if the fluidized bed were at atmospheric pressure, and to merely a pair of semi-detached bijou residences if eight or even sixteen atmospheres of supercharging were utilized.

The fluidized bed has other prospects too. The pressurized or supercharged version could be utilized with a gas turbine, perhaps in a combined gas/steam turbine cycle of high efficiency. Another prospect is to use the gas from the fluidized bed boiler as a feedstock – presumably after extracting some of its heat energy – for any of a variety of possible processes. This idea of multiple use of the gaseous output is called the 'coalplex'; the output could be a synthetic natural gas – which means old-fashioned town's gas with a new look – or with hydrogenation, a synthetic crude oil. In simple terms, it could be the starting point for a 'coal refinery'. If that proved to be the case, it could be the starting point too for an economic resurgence of coal.

Energy on current account

A distinction has been made between the 'capital account' energy of fossil fuels and nuclear transmutation and the 'current account' of solar energy and the cycles of weather that derive from it. There is only one source of current account energy that is of any great practical importance today, and that is hydro-electric power. Its future holds promise of greater mastery over the brutal conditions of nature, in the ability to build dams at difficult sites, to handle flood waters and to tunnel quickly and confidently, for example. Improved transmission would increase the prospects for remote sites.

Another fifty to a hundred years could see the world's major hydro-electric sites exploited, including not only rivers, but evaporative sites such as the Qattara Depression in Egypt, where the sea would be diverted through turbines into the depression at such a rate that the depression would be depleted by evaporation and would not fill up. The Dead Sea is another site of this kind. Tidal power offers further prospects, but it suits our narrative to discuss the wind and the sun first.

The wind is so obvious and so available; indeed windmills are so traditional. Why is the kinetic energy of the wind not extensively utilized in this power-hungry age? For the answer, who better to go to than a firm that sells windmills. A note published in the journal which I edit about some research on a new form of windmill provoked the following letter from Mr T. J. F. Benjamin of H. J. Godwin Ltd, of Quenington, Gloucestershire, which in this context can stand on its own without further explanation.

Sir,

I was intrigued by your article 'A Practical Wind Turbine', as my company has all but ceased to manufacture windmills, for the very practical reasons that demand is almost non-existent. Perhaps I can add some experienced observations for enthusiastic wind power engineers which I hope might orientate their thinking.

Reference was made to the vertical high speed turbine eliminating the heavy gearing required to keep the mill into the wind; alas this is the simplest part. Reference to a post mill or modern pumping windmills will indicate that the horizontal fan shaft rolling round its own pinion is a simple and satisfactory way of maintaining direction whilst effecting the necessary radial and axial shaft support and speed reduction. In our own range of Hercules mills, regulation is achieved by a fan tail and the mill can be made fully automatic, turning itself out of the wind when pumping has been completed.

I will avoid going over the ground that has already been covered in much detail on wind machines, but it is not surprising that the designers of this new turbine found that their design had been done before and probably before that.

There are three governing factors which make or break all wind machinery (and you may include sailing craft):

1 Capital cost is high for the power generated.
2 They are of very limited flexibility.
3 The wind is variable and fickle.

Capital cost can be largely off-set by low running costs and we must compare these devices with say diesel engines in this respect. A diesel-driven pump however, is more reliable and is more flexible both as a machine itself and in that the prime mover is one of a mass-produced range used for many different purposes. Its capital cost reflects this as does its fuel cost.

The major stumbling block is the fickleness of the wind. It will be noted that the most successful wind machine (other than the sail) is the wind pump and it is extremely useful to analyse why.

These devices almost invariably have a reciprocating pump and not a centrifugal pump. A centrifugal pump develops its performance as the square of the speed; hence variations in speed cause either a rapid fall in performance or heavily surging power consumption. A positive displacement pump is not dependent on speed for its head, and always pumps independently of wind speed.

Relate this now to electricity generation and it is not difficult to see how a flukey wind can play havoc with the generated voltage. This has the further effect of loading any controlling switchgear excessively.

The obvious reaction is to use variable pitch, but whilst this is an improvement, one cannot steepen the pitch enough in a calm, and it takes a wind of 15–20 m.p.h. to make the concept viable in the first place. Variable pitch also means higher capital cost with higher maintenance.

The second part of the problem is storage, and it can be seen why the reciprocating pump/windmill combination has remained relatively successful. The pump will supply something in all but the calmest weather to the reservoir. Energy can be stored in a high reservoir, often on ground unsuitable for cultivation, requiring minimal maintenance and low capital cost. Extending the analogy to electricity generation, the success of wind generation depends on improvements in storage. Not one design for electrical storage currently in development even approaches the simplicity of a reservoir of water.

The windmill in the United Kingdom is now largely replaced by our range of electro-submersible centrifugal pumps. We have a number of mills working which we still service, and we supply a number abroad. Certain areas of the world, the Sudan, South Africa, Australia, are still very active in this field and we also supply these countries with our K reciprocating heads, diesel-driven, for when the wind fails. Our activities now relate mainly to electric and diesel-driven pumps which we export throughout the world.

If it became possible to develop storage techniques for the captured power, then wind power could again become viable. In this respect there are many parallels that can be drawn with electric cars. What is also required I fear is not just another form of turbine but a considerable rise in the cost of the alternative fossilised or atomic fuels. Electricity and the diesel engine are such tough little workers.

As a company of 90 years standing and considerable experience with windmills, we are getting increasing requests for information on wind generators. We are naturally most keen to use this experience but for the reasons outlined above, it will be obvious that there are inherent drawbacks which are quite major to the whole concept.

On a lighter note I would recommend readers to stop in the Cotswolds and view some of the mills there. A mill near Tetbury, of our make has a 16 ft diameter sail (no Mr Metricator, not 4·7 m) and is a sight of silent power, humming round in a stiff breeze with the pitch-pine pump-rods gently clanking their metal forks and guides on each steady stroke. On a summer's day it is a memorable sight for an engineer. Alas, there is an electro-submersible in case the wind fails.

Fickleness is a costly attribute, and others could repeat similar arguments. For example experiments with large windmills mounted on tall towers at windy sites, and intended specifically for the generation of electricity, were carried out by the British electricity people soon after the war. But an electricity undertaking wants firm power, and an auxiliary source that has to be backed up has a limited value.

Solar energy

I once attended a publicity function at which an American maker of solar cells had rigged up a delightful electric car of Edwardian vintage with solar cells on its roof. The solar cells drove the car directly or charged its batteries. It was a foggy November day, so we all retired for drinks and went home on the underground. Other climates have reason to be more optimistic; for example solar energy is utilized domestically in Arizona. At Odeillo in the French Pyrenees there is an 'optical concentrator' that can yield over a megawatt of heat energy and reaches a temperature of over 3,000 deg.C.

The best review of the prospects for solar energy that I have read is by Professor Martin Wolf* who relates the four principal methods of utilization – through meteorological or geophysical effects, biological effects, direct use for heating, and conversion into electricity – to an assessment of the general picture of the world energy economy and its growth as he sees it. He concludes that 5 per cent of the U.S. demand for energy in the year 2000 could be met by utilizing solar energy, and 25 to 30 per cent by 2020. In summary, solar energy is the inexhaustible source, free even from the objection to nuclear fusion that it upsets the terrestrial energy balance, and is the philosopher's choice of source for the future welfare of humanity. But at present it is more of a scientific plaything than a practical source.

Tides

Before the days of electricity, tide mills were used at convenient spots, mainly to grind corn. There is one left in Britain, in Suffolk. The miller

* Professor Martin Wolf, 'Solar energy: an endless supply', University of Pennsylvania. *Consulting Engineer* (*U.S.*), March 1973.

was prepared to work when the tide was right, about an hour later each day, and doing more work at spring tides (once a fortnight) than at neaps. An electricity grid is less flexible; it wants power at peak hours. For years, the fact that tidal energy is not continuous, and that its availability varies daily with time and over fortnightly cycles by amount, were sufficient reasons to prevent its utilization. The Severn barrage was designed seriously several times, and in the late 1940s justified solely on the grounds of the coal it would save (virtually the same economic case as for windmills) but in those days of shortages of all resources, there were more urgent needs for the large capital sum that would have been spent to build it.

One early attempt to overcome the characteristics of the tides and produce a nearly even flow of energy from them is interesting as a historical curiosity. In the late 1920s a small tidal power plant was installed in a dock at Bristol by a former chief engineer of the Russian navy, Mr Shishkoff, who had come to Britain before the revolution with the aim of developing a gyroscopic torpedo for the British Admiralty. In the Shishkoff hydro-thermal generating plant, as it was called, a conventional water turbine drives a generator when the tide falls sufficiently for a head to be available across the dock gate. As the head increases and the turbine begins to generate more power than a limiting value, it begins to drive a dynamometer as well as the generator. Heat is generated in the dynamometer from which steam is raised. The steam is passed to the reservoir, and stored. As the tide passes the peak head, the dynamometer comes off load, and later, the output of the turbine falls below the minimum limiting value. At this point, the steam is fed to a steam turbine mounted on the same shaft as the other two machines, which then drives the generator. The quantity of steam stored in the reservoir is sufficient to keep the generator going until the tide has again reached the stage in its cycle where the water turbine takes over. I think description of that unique group of machines should remain unadorned by any comment.

La houille bleue

The first tidal power station, and up to now the only one of significance (there are one or two tiddlers in Russia), utilizes the estuary of the River Rance in Normandy, close to St Malo and Dinard (Plate 28). The success of the Rance scheme depended in an interesting manner on the resolution of three interconnected problems. First, a civil engineering one – the closure. When a tidal estuary is dammed, as the gap in the dam narrows, the flow through it at each tide increases in velocity and in its destructive force. Thus a tidal closure is necessarily

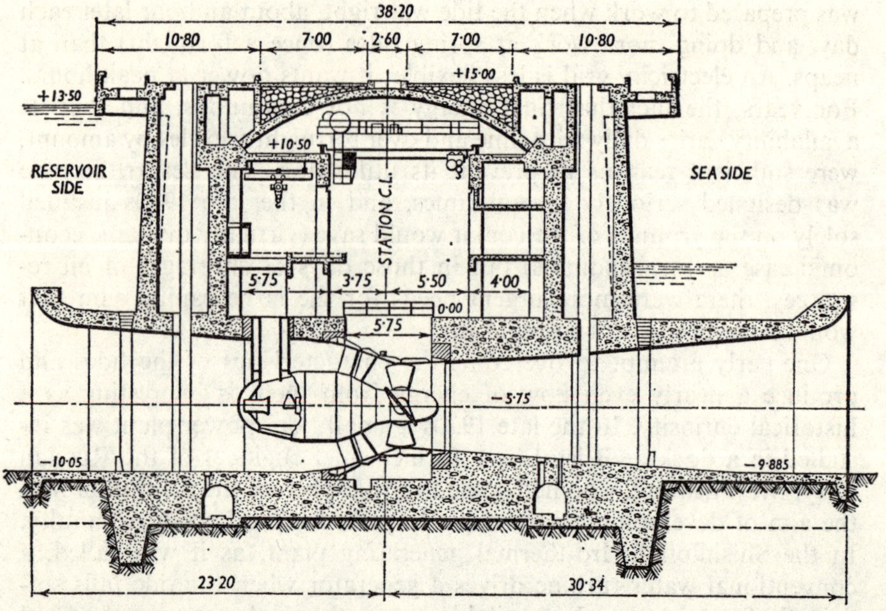

Fig. 14 Cross section of the Rance tidal power station, showing one of the bulb sets. The nominal capacity of the set is 10 MW, and there are twenty-four sets installed.

hazardous; at the Rance, it was pushed beyond previously experienced limits so that a reservoir of significant size, and hence output of power, would result.

The cycles of output at varying hours and of varying amounts were attacked in two ways – problems two and three respectively. An entirely new type of turbine set was evolved, the bulb set (Fig. 14), which could operate in both directions of flow – that is, with flow either into or out of the reservoir – and over a range of heads either as a turbine, as a pump, or as an open sluice. A mathematical research into the possible variations in output which could be achieved with sets of this kind was undertaken (actually the sets were specified from the maths) by Electricité de France's consultant, after whom the variations were named Gibrat cycles.

For simplicity, the solution of these problems can be reduced to this: the bulb sets went part, but not all of the way towards achieving a regular output of power from the natural variations of the tides. But which parts of the variations were eliminated was a matter of choice – again within limits. For example, water could be pumped into or out of the reservoir at certain times to fill it higher, or empty it lower, so increasing the power generated at certain specified times later; without pumping, subsequent generation would be different in amount and timing. The Gibrat cycles consisted of analysis of the possible future outputs. The output selected would depend on the needs of the electricity grid. By feeding the immediate requirements of the grid into the analysis, the most needed possible output would be selected and the power station operated accordingly. If peak power were needed, it could be maximized at the expense of total output, and so on with almost endless variation. Different monetary values were assigned to different classes of output, to arrive at an economic evaluation. Thus flexibility in operation not as complete as for a well-reservoired hydroelectric scheme, but very much greater than for a tidal scheme as conceived before, was achieved. The economic value of the scheme was considerably greater than merely the saving in fuel credited to the earlier Severn barrage proposal.

In spite of its engineering elegance and its success, the innovation has not been repeated. Advances in conventional thermal generation overtook it, just as they overtook nuclear power and many marginal hydroelectric proposals in the sixties. Modern trends – but not yet translated into an actual scheme – have gone in another direction. The philosophy is: build the scheme very cheaply, even if some of its output, and in particular the flexibility of the Rance scheme, is lost. Thus a cheap dam that leaks would be sufficient; if acceptable structurally, the leaks are of no consequence because of the low head and the enormous quantities of water involved. The dam would be built to about low-tide level and the closure made by a method recently evolved in Holland, by floating out caissons and placing them along the top of it. The caissons have gates that are left open, and when they are all in position, the gates are all closed at slack water – very neat. A very cheap, combined turbine-generator machine (recently developed) would be installed in each caisson; the moving parts of the generator are on the turbine itself, a technique made possible by the efficacy of modern sealing methods. All very cheap and nasty compared with the elegance of the Rance, but still only competitive by a slender margin if you are lucky. Such a scheme has been prepared in outline for the Severn – for example – a site that every proponent of tidal power casts his eye over – but there is yet another reason against building it.

Really large outputs are sought these days to make worthwhile contributions to the big demands on the national grids. A second stage of development after the Rance was proposed in Normandy, much larger and actually in the sea (Fig. 15), as shown in the sketch. Ownership of the Minquiers and Ecrehous islets was contested and the International

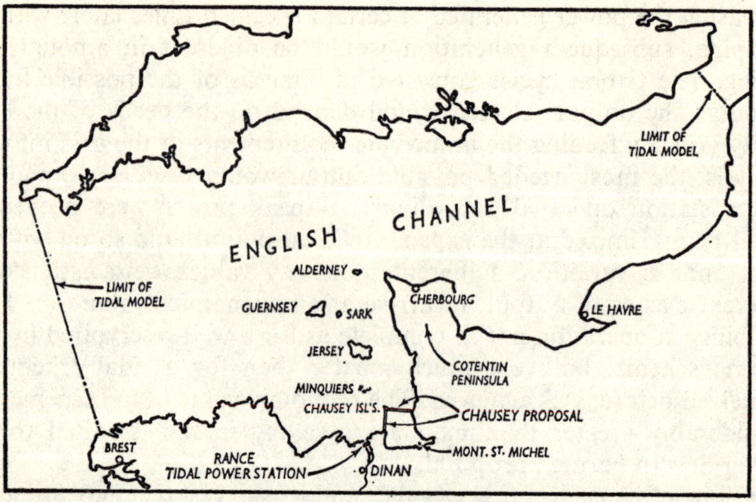

Fig. 15 Sketch of the Mont St Michel proposal for tidal power. Its output was investigated on tidal models, and originally larger stages of impounding extending to the Channel Isles were considered.

Court at the Hague ruled in favour of the Channel Islands (Britain) and against France, but by then the case had become a dead issue. There are not many sites where the amplitude of the tides is sufficient to make them attractive for tidal power, and this was one. But research indicated that construction of the scheme – a square of sea would have been impounded at the corner of the French coast by two dams (about 20 and 15 kilometres in length respectively, both built between the Chausey Islands and the mainland) – would itself have reduced the amplitude of the tides by about 15 per cent, so reducing the output which had been estimated.

In estuaries and near constricted coasts, the tides are resonant in character. The converging shape of an estuary causes the flood tide to rise progressively higher as it penetrates farther up the estuary. If a dam is placed across the estuary, this effect is destroyed. By damming the square of sea outside St Malo, the resonant effect, present in nature, would have been reduced. In a scheme of the size proposed (the Rance scheme is too small to be affected) the reduction of output was sufficient to upset the economic balance and make the scheme rather too costly. The strategy of French electricity supply was also changing, and the big scheme, after delay, was seen as an isolated exercise with a disproportionate demand for capital in relation to the French economy as a whole.

A big scheme on the Severn would also have an output less than would be estimated from the amplitude of the natural tidal régime. The Severn and the Rance are probably the best sites in Europe. On the other side of the Atlantic there are potentially better sites in the Bay of Fundy where the tidal range is the largest in the world. These sites too have been repeatedly investigated but not exploited.

That in outline indicates the main stream of tidal power technology, but there are other possibilities as yet untried. One intriguing suggestion is to make use of a device which has been obsolete since the industrial revolution, the hydraulic ram. By using the tide to operate a series of rams water could be lifted to a reservoir 30 or 40 metres above sea level. The inconveniently low head of the tides could thus be converted into a useful one, and the reservoir would function like a pumped storage scheme – which of course it would be – without temporal limitation.

Another possibility is to use not the change in level of the sea, but the sea current itself, whether tidal or an ocean current. This could be done with simple, cheap, combined turbine-alternators like the machine just described, spotted strategically about the sea bed just like windmills, and serviced by a rig like those commonly used for offshore work. Alternatively, larger versions of such machines could each be sited at the throat of a big submarine venturi. It is reported that Russia plans to build a power station to utilize the tidal currents in Penzhina Bay in the far east, where the bay narrows to 32 kilometres between Kamchatka and the mainland. A different approach from the one described here is favoured, and the power station would float.

The Channel Dam

The sea currents are sources of kinetic energy, but also of thermal energy if the moving water is warm, like the Gulf Stream. The waters on the western approaches to Britain are warm because of the influence of the circulation typified by the Gulf Stream. The mild winter climate of the western coasts of Europe is due to south-westerly winds, which have been warmed by picking up heat from the sea over which they have blown. Thus warm water and mild winds derived from it give a climate up to 10 deg.C warmer than at the same latitude on the other side of the Atlantic. Perhaps this is nature's example of thermal pollution?

Can this effect, as presented by nature, be improved? Yes, it seems. The principles for doing so have been explained by a French engineer, Camille Rougeron. I translated and published M. Rougeron's work on this subject nearly ten years ago, when the Channel Tunnel controversy was topical (sic). His ideas were resisted with some clamour, and I even heard them described as sinister.

Here then is the principle. The influence of the Gulf Stream creates a sea of warm water along the western coasts of Europe. In summer, the influence of these waters is secondary to that of the sun, and isotherms of average temperature run roughly east to west – it gets colder as one goes north. In winter, however, their influence is felt most strongly on the west coast, and attenuates as one penetrates the continent towards the east. In Britain, the isotherms run north to south – it gets colder as one goes east. The milder winter climates are in Brittany, Cornwall and the west coast of the Highlands, where there is less frost than at Dover.

Mild winters are also enjoyed in the Channel Islands and the Cotentin peninsula, but farther east the coasts of the narrow seas are less mild. The sea temperatures during winter show similar variations. The temperature of the southern North Sea drops in winter more sharply than in the western half of the English Channel, because warm, oceanic waters from the west barely penetrate to it. The tides in the Strait of Dover, first rising up the Channel, bring a current from west to east through the Strait, and then, rising southward down the east coast of England, a current from east to west through the Strait. There is thus an oscillation of water, but only a small net flow to the east through the Strait. Taking the triangle of the North Sea with its apex at the Strait and its base the latitude of Yarmouth or thereabouts, the volume of water thus defined is about half renewed in a winter.

M. Rougeron points out that if the current from west to east could be encouraged, and the east to west one retarded, to the extent of say four renewals in one winter of this triangle of sea, then its temperature, and the temperature of the water in the eastern half of the English Channel, would be increased and would be little less than that now observed in the western Channel. This warmer water would attenuate the climate of late autumn and winter in the southern part of England and in northern France and the low countries. Nature's régime in Cornwall and Cotentin would extend eastwards along the Channel, in proportion to the increased penetration of warmer water, and milder weather to the extent of maximum increases of average temperature of about 2 deg.C in the late autumn would be experienced. Effectively, two thermal reservoirs would be replenished by the renewal of water from the west, the eastern half of the Channel and the southern triangle of the North Sea. Winds blowing over them would pick up heat. The precise effects can be predicted by tracing the paths of typical weather systems as they travel over the area and noting how the winds would be warmed at any stage.

How would renewal of the North Sea water be achieved? Several methods offer possibilities. The simplest to follow physically would be a set of flap gates, as big as the Strait of Dover, which stopped westward flow, but that can be considered as a demonstration rather than a proposal. Any proposal for partial closure would need to be investigated with a tidal model. Training walls, perhaps part of a combined bridge-tunnel crossing along the lines of the Chesapeake Bay bridge-tunnels, perhaps with flap gates at some places, could be designed to promote westward flow and retard eastward flow. This approach could be combined with an installation of pump-turbine machines working on the sea currents, either taking energy from the currents or giving the current a boost in velocity. Alternatively, these machines might do the job on their own. They would be of either of the two types just described, working on lines similar to the Gibrat cycles of the Rance scheme. When tidal currents were slow, except at peak hours on the grid, they would pump towards the east; when tidal currents were high they would generate. Energy is proportional to the square of velocity, and there would be a net output of power, larger in summer when no pumping would be needed. If a tidal surge were forecast, such as happened in 1953, the machines would pump westwards, to reduce its peak. The size of the installation could be modest at first, increasing with experience, so the effects of the scheme could be sampled on a small scale, and then built up.

Economic justification of the proposal is interesting. The power itself would seem to be about as attractive as more conventional tidal power proposals – that is, marginal in the eyes of electricity utilities. Multiple-purpose design, combining a channel crossing as an additional objective, offers greater possibilities, unless one happens to have landed a fat concession for the tunnel already, when any other idea becomes positively embarrassing. In either case the greatest savings would be of fuel in autumn and winter, throughout the areas where the climate would be attenuated; this saving would be so large as to overbalance the other considerations, but it would accrue neither to the promoter nor to any one pocket; in fact it would embarrass the purveyors of fuel by reducing their sales. Perhaps its greatest advantage – the softer comfort of a long Indian summer – cannot be entered even in the fictitious accounts of an economist's books.

To work these principles to the stage of calling the Channel dam a scheme would require a deal of investigation, but none that was out of proportion with what is accepted as necessary for other engineering proposals. A similar approach is possible in similar latitudes on coasts facing warm currents, and the Russians have a similar proposal for the Sea of Okhotsk; its economic benefits could hardly compare with the Channel dam however, which would serve the golden triangle of the E.E.C.

The principle utilized is to divert to better advantage the heat stored the preceding summer in the warm waters of the tropical ocean, and then transported by natural ocean currents to colder regions; heat transfer to the land needs the winds as the thermodynamic working agent. The Rougeron proposal and its Russian parallel are modest in size by comparison with the climates in which they would operate. The general picture of European climate would be modified to barely a marginal extent, even if the idea were pushed to its limits, but with great benefit nevertheless. Extra moisture would be picked up by the winds, but with imperceptible extra rainfall. The major forces of climate would be untouched, and the autumnal attenuation would be just as if produced by some natural idiosyncracy of geography. Large bodies of water attenuate climate everywhere; maximum temperatures over land tend to lag about a month behind the sun; in the sea the lag is about two months. Students of thermodynamics would have expected it so, but by replenishing the warm water, the lag is extended over a still longer period.

Can climate be modified?

The factors that control climate are not understood. Ignorant engineers should not therefore meddle with it. That message has been stated loud and clear by at least one student of oceanology. Such simplistic generalizations tend to be influenced by a competitive spirit between professions, and the scientist may use as his ninepins proposals that might be rather *outré* amongst engineers. These exchanges make good copy for journalists and authors, but are better replaced in the world at large by the sort of approach that I explained and praised in relation to the multiple-purpose development of river valleys. Each specialist has an important role, but unless and until their roles are properly related to each other, and each related to the basic aims and strategy of the development, then lopsided projects will result.

Nevertheless the message, within limits, has validity should any major change in climate be proposed. The Rougeron scheme and its parallels are not in this category. Effects on a similar, maybe smaller scale, have already resulted from engineering schemes, e.g. bringing large tracts of desert under irrigation or making reservoirs the size of those at Kariba or Aswan.

On the other hand, idiosyncrasies of geomorphology seem to have profound effects. The Gulf Stream is a good example. It seems to have been formed by the peculiar accident of a natural venturi throat, thus:

> The Equatorial currents are forced to move northwards, and skirting around the islands, the piled up water in the Caribbean Sea

forces itself out in a gigantic squirt through the fifty-mile-wide Florida Straits between Cuba and Miami. At the narrowest part of the strait, the Narrows of Bimini, the flow of the current reaches to the sea bed at a depth of 1,500 feet and passes through at the rate of ten cubic miles per hour. The speed of the current . . . reaches seven miles per hour . . . on average it carries more than twenty-five times as much water as all the rivers of the world added together. This vast jet of water, fifty miles wide and a quarter of a mile deep, aims itself northward along the coast of Florida . . . it maintains its entity for some thousands of miles . . .*

Various methods of making use of the Gulf Stream have been suggested – Mr Gaskell's book gives the details – but they are somewhat fragmentary in the context of understanding the general circulation of water in the oceans, the reasons why the Gulf Stream and other currents exist, and the effects which they have. Armed with this knowledge, the possibilities offered by 'meddling' with the ocean circulation become a fascinating field of study. A minor, attenuating effect at the tail of the current, like the Rougeron proposal, could be followed, after gaining experience with it, with some more ambitious attempt for man's benefit, perhaps leading eventually to new geographical zones of salubrious, verdant lands warmed by man-made gulf streams, which, as a minor benefit aside from the creation of a pleasant environment, would cause the inhabitants of these places to use less fuel to heat their buildings.

Geothermal heat

Speculation about climatic modification leads the narrative somewhat deviously back to nuclear explosions, and thence to geothermal heat. M. Rougeron's proposal, as I have described it, is a sort of aside in his book on the fusion explosion.† He makes other suggestions for climatic modification which depend on the use of such explosions. For example the St Lawrence River, he suggests, could be kept free of ice in winter. A deep crater on the upper part of the river, formed initially by a fusion explosion, would tap geothermal heat by heat exchange with the water with which it was filled, and so would function as a thermal reservoir – the water being at 4 deg.C – with capacity sufficient to keep the river free of ice throughout the winter. His book was written nearly twenty years ago, and even with the strength of the 'naturalists' today, I doubt whether humanity could become less friendly towards liberal attitudes

* T. F. Gaskell, *The Gulf Stream*, London, Cassell.

† Camille Rougeron, *Les Applications de l'explosion thermonucléaire*, Paris, Editions Berger-Levrault.

to fusion explosions than it was at that time. The St Lawrence's ice is likely to remain sacred. Even so, Rougeron's ideas have been investigated in practice by the American Plowshare programme, but Plowshare itself seems to be dying now. Explosions for large works of construction – the proposed sea-level canal at Panama for example – seem to be out of favour. Yet the Russians have used conventional explosives in a highly dramatic feat in which one large explosion formed a dam of considerable size (Baipaza dam on the Vakhsh river in Tadjikistan – 60 m high, 1·2 million cubic metres volume), by blowing out the sides of the valley with nearly 2,000 tonnes of explosive.

One of Rougeron's suggestions was that an underground fusion explosion could be used to form a cavity, well below the earth's surface. The energy liberated by this explosion would be supplemented several times over by the geothermal heat stored in the debris in the cavity, which could be extracted by circulating steam to a power station on the surface. He saw in this an abundant and cheap source of power. Plowshare has detonated such subterranean explosions, but reports that steam leaked out of the cavity to the surface through fissures, and the idea has apparently been abandoned. According to newspaper reports, the most recent U.S. test at Amchitka Island – which was bitterly opposed – was intended to be a subterranean explosion but it broke through to the surface. The results are secret, and clearly I cannot vouch for the reliability of the report or for any significance that can be read into it.

Geothermal energy exists as an enormous source. Its use, lacking a Rougeron style of approach, is very limited, and is restricted today to sites where steam or hot water spout naturally from the earth's interior. At Lardarello in Italy such a site has been utilized for power generation for many years. There are now also sites utilized in New Zealand; one on the U.S./Mexico border has also been under development, and in Iceland geothermal hot water is used for domestic heating. The total output from this source is, however, very small.

As for the peaceful use of nuclear explosions, in the U.S.A. the best prospects are considered to lie in what is called gas stimulation. An explosion is detonated which makes possible an additional yield of gas from a field that has been nearly exhausted. Testing of this technique is active in the U.S.A. In Russia similar tests have been carried out on oil-well stimulation. The explosion may (*a*) shatter rock and increase its permeability, (*b*) break down a barrier and so expose the field to natural high-pressure water, (*c*) reduce the viscosity of oil in carbonate rock (by dissolved CO_2), and (*d*) break up unwanted pressure-relief channels. The Russians also used a nuclear explosion to extinguish a fire that had been burning in the Bukhara gas field for about three years, consuming enormous quantities of gas.

By contrast, apparently the French view is that underground storage is the most promising application. A 'nuclear chimney', and a space formed beneath a natural dome of salt, have been experimented with. A fuel – gas or oil – could be stored.

There have been reports of the use of nuclear explosions for construction in Russia. It would seem logical that the success of the Baipaza dam would favour such a trend. In 1971 a nuclear explosion was detected close to the alignment of the Pechora–Volga canal. The Pechora is a river that flows northwards, to the Arctic coast. The canal would divert its water southwards down the Volga, through the chain of power stations in that river, for eventual use for irrigation.

13 Entropy in the crystal ball

We have gone full circle, and returned to the theme of Chapter 1. There are some interesting long-term perspectives ahead, slipping momentarily into view from behind the clouds in the crystal ball. The shorter term, covering roughly the next doubling of demand for energy (if the history of its growth continues as before) and extending into the mid-1980s, has already been under the eyes of the planners and forecasters, but even so there could be some surprises. (See Postscript.)

The longer term, after the year 2000, is more speculative, but there is a common pattern amongst the studies and prophecies – of which there seem to be too many – the main lines of which are thus: fossil fuels are man's main source of energy; they are finite in quantity and cannot be renewed. Given that population and consumption per head are both rising, the time is not far distant when they will be exhausted. For example, world energy consumption in 1968 was about 6×10^9 tonnes of coal equivalent, and in 2000 will be about 30×10^9, a fivefold increase.* The conclusion has thus been drawn:†

> If these substances continue to be used principally for their energy content, and if they continue to supply the bulk of the world's energy requirements, the time required to exhaust the middle 80 per cent of the ultimate resources of the members of the petroleum family – crude oil, natural gas and natural-gas-liquids, tar-sand oil and oil shales – will probably be only about a century. Under similar conditions, the time required to exhaust the middle 80 per cent of the world's coal resources would be about 300–400 years, but only 100–200 years if coal is used as the main energy source.

Many contemporary pundits see oil reaching its maximum use by the end of the century. It would then decline in importance and there would be a resurgence of importance for coal, perhaps with quite novel methods of exploitation and use, for a century or two. If this forecast is approximately correct, the entire history of the use of oil and coal could be roughly as shown in Figs. 16a and b. Ultimately, nuclear sources of energy would replace fossil fuels. A variant on this theme – but slightly

* Sir Harold Hartley, *New Scientist*, 13 November 1969; see table 10 for conversion factors.

† Dr Hubbert, 'Resources and man'. National Academy of Sciences and National Research Council (U.S.A.), 1969.

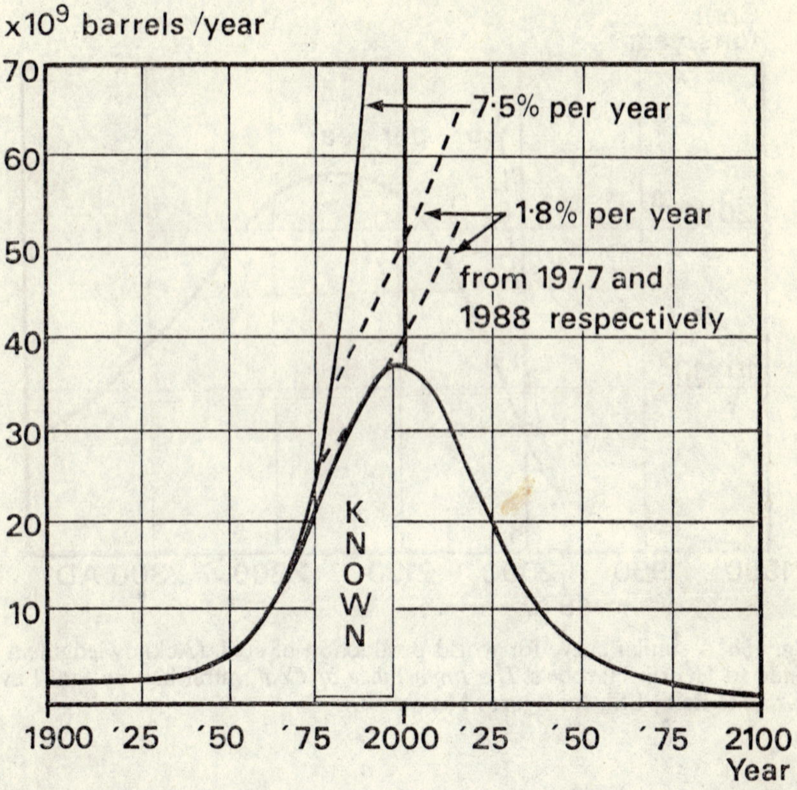

Fig. 16a A possible historical cycle of world production of oil. The area marked 'known' shows the extent of proved reserves. The area under the curve corresponds to a total world production of 2100 × 10^9 barrels (approximately 0·5 × 10^{12} tonnes coal equivalent) and is twenty-one times the estimate of oil already consumed (Table 11). If production increased by 7·5 per cent a year, the solid line indicates the quantities that would result; similarly, the dotted lines indicate 1·8 per cent a year increase, starting in 1977 and 1980 respectively.

more complicated – is shown in Fig. 17 from a study made by Erik Grafström, president of the Swedish State Power Board.

Immediately ahead

In the industrialized countries, trends are well defined and amply documented. The outlook is that growth of energy usage will continue and that electricity will continue to increase its share of the total. Nuclear power will grow faster than electricity as a whole, perhaps to a substantial part (half in the U.S.A. according to Dr Sporn) of electricity's

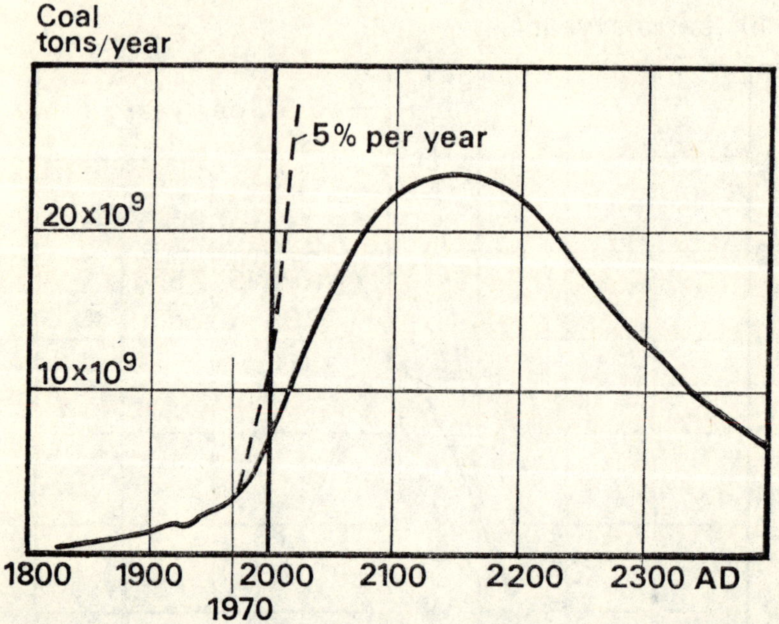

Fig. 16b A similar curve for world production of coal. (Acknowledgment is made to Henrik Harboe's *The Importance of Coal*, published by Stal-Laval [Great Britain] Ltd, for figures 16a and b.)

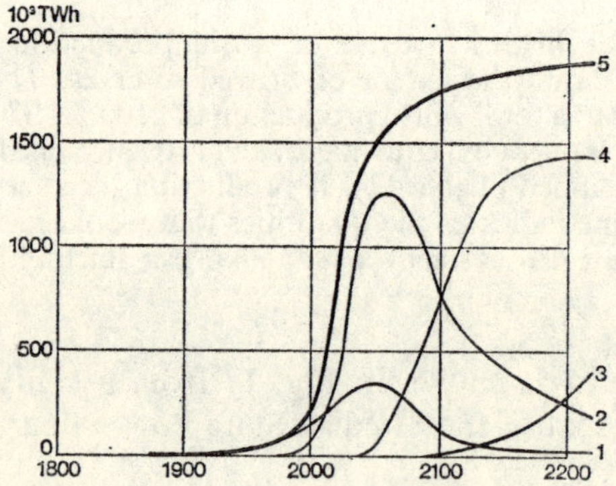

Fig. 17 Growth of world energy demand as seen by Eric Grafström:

1. Fossil fuels
2. Fission
3. Solar energy
4. Fusion
5. Total. The levelling off towards AD 2200 is about 60 per cent of the 'levelling off' figure given in the text.

Figs. 18, 19 and 20 These three graphs are Swedish studies of future energy patterns. The first shows how demand is beginning to outrun hydro-electric capacity in Swedish electricity supply. The second is electricity consumption by category as projected in Sweden. The third is an estimate of future world capacity for generating electricity.

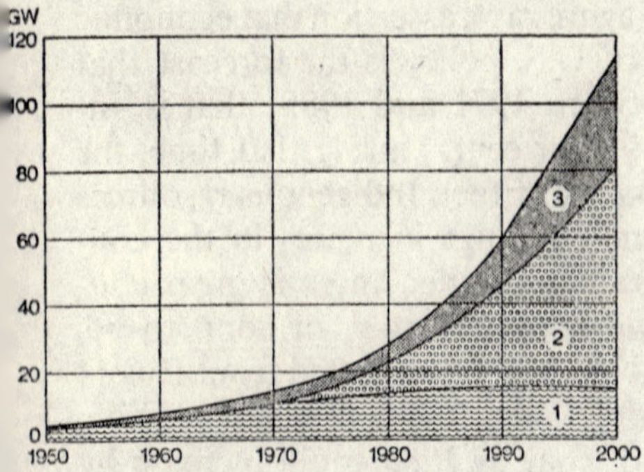

Swedish generating capacity: 1. hydro; 2. nuclear; 3. thermal (fossil).

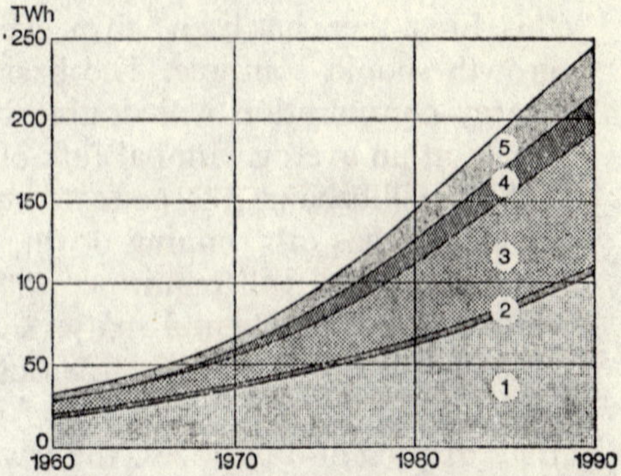

Swedish electricity consumption: 1. industry; 2. transport; 3. domestic and commercial; excluding 4. electrical heating; 5. losses.

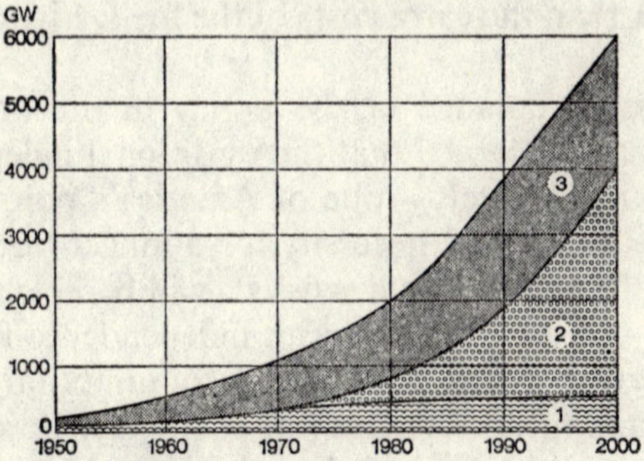

World generating capacity: 1. hydro; 2. nuclear; 3. thermal (fossil).

share by the turn of the century. These forecasts assume there will not be rapid, unexpected innovation. Past experience tends to confound that view, but today technology is so complex that innovation has become ponderous. The choice of view is open.

The dominant factor in recent decades has been growth in the consumption of oil; at present 80 per cent of the proved reserves, mainly in the Middle East, are controlled by OPEC (Organization of Petroleum Exporting Countries). The days of cheap oil may be numbered; politically-inspired haggling will outweigh statistical appreciation of natural resources in the future pattern of usage. Combine this fact with the predictions of mounting consumption, and it is clear why there are voices in the west which advocate national energy policies and an organization of consumer countries. (See Postscript.)

By far the largest user of energy is the United States, whose policy is this. First, there has been a firm, even aggressive, assertion that economic growth should continue. The basis of U.S. policy is the forecast that energy consumption will double between 1971 and 1985; that is, increase at an average annual rate of 4·2 per cent. During that time, the U.S.A. will become a large-scale importer of fuel. Indigenous resources of natural gas are running down, and the huge increases in the consumption of oil will require considerable imports. There is great concern over environmental matters, and measures now, or soon to be, enforced will add extra costs. Natural gas will be imported from Russia and Algeria, oil mainly from the Middle East. It is considered possible that, at present-day prices, the cost of coal by 1985 could increase by 30 per cent, of oil by 100 per cent and of gas by 200 per cent. If one recalls the principles explained earlier, justification for this policy becomes understandable. The extra demand for these fuels will be created by increasing economic activity of all kinds, which will generate wealth generously greater than the extra costs of the fuels. Living standards will continue to rise.

Necessity forces acceptance of the policy in the opinion of many Americans, who are concerned that the traditional independence of the U.S.A. from imports of fuels – one of America's points of strength – should be eroded. They see a necessity for a longer-term policy which, while permitting the trends just discussed, will be able to reverse them in the 1980s, and re-establish American independence from imports of fuels, even at the greatly increased rates of consumption which will then be needed. Ways and means of achieving that longer-term policy should be worked out now, it is asserted. For the immediate future there are plans on a great scale to handle the growing flow of imports – liquefied natural gas and oil – to increase indigenous output, for example of coal, and for measures such as making gas from coal. As I write (1973), a bill has passed the Senate and been sent to the House, to make a three-man energy policy council responsible for the centralization, collection and analysis of energy information . . . to co-ordinate energy activities of all existing agencies and prepare a long-range plan for energy utilization.

Imports on this scale by a wealthy nation will inevitably push up prices and make buying more competitive for other importing nations, which in the next decade will include virtually all of Western Europe.

Britain has traditionally relied on Middle East oil, and has been an importer of fuels for decades; the security of an indigenous energy economy, so prized in the U.S., has been a luxury beyond Britain's ken. There is, however, so much change taking place, that traditional attitudes and patterns of supply are no longer much of a guide. In the new European community, Britain is preponderantly the largest producer of coal. The other countries of the community import larger percentages of their fuels than Britain (on average 63 per cent, compared with 45 per cent for 1970). In 1925, Western Europe imported only 2 per cent of its energy. Table 7 shows that Western Europe will be seeking to import in 1985 more oil than the U.S.A. and will be less self-sufficient than the U.S.A. in energy. Exploitation of new sources of oil in Alaska and the North Sea will reduce, perhaps only marginally, the dominance of the Middle East as the world's major source of oil during that period.

Britain thus seems to be rather more favourably placed than the average E.E.C. country. The energy economy of Britain is now based on four fuels: coal, natural gas, oil and nuclear fuel. Discoveries of natural gas and oil under the North Sea are great assets. The best estimates that can be given are that by the 1980s, about 70 to 100 million tonnes of oil will be won by Britain annually from North Sea resources (the higher figure is nearly equal to present consumption), together with all the national demand for gas. The National Coal Board is alive to the opportunity which current circumstances present to it. Coal is too dear, but if its price can be held while the prices of other fuels increase, there will be a good market for Britain's coal throughout the E.E.C.

An energy commission, or even a policy?

Lord Robens* recently pleaded for the establishment of an energy commission for Britain. His view is that a positive energy policy is needed; it would have as its basic points 'complementary and rational utilization of all this country's indigenous fuel resources' and 'the importance of secure energy supplies is out of all proportion to the cost'. Hence the policy would resolve who would carry spare capacity and stocks (which is detrimental commercially but desirable strategically, according to Lord Robens), would limit competition between indigenous fuels, and while not divorcing the fuel industry from commercial pressures, would reassess commercial rules to give greater weight to long-term considerations. The Civil Service is inadequate to do all that, in Lord Robens' view, and an energy commission would have to be set up.

* First Potterton lecture, Institution of Heating and Ventilating Engineers, 1973.

Other exponents of an energy commission seem to stress other arguments which are generally variants of the total energy theme, i.e. that many purposes can be served by one fuel, maybe in one installation, and just selling the fuel competitively is not sufficient. This argument particularly criticizes the Electricity Council's single aim of selling electricity, to the neglect of district heating and other techniques of utilizing heat now wasted in electricity generation.

The point of view that market forces are the best spur to improvement, efficiency, and cheapness has many advocates. Rather significantly, they seem to include many of those with long experience and heavy responsibility in the energy industries. Lord Hinton, in discussing future energy needs, listed the learned committees which had been appointed to foresee and assess the future of energy in recent decades – British and E.E.C. Their reports had one thing in common – they had all proved, after two or three years, to be useless; reversals of policy had followed them, and at least one of them had been positively harmful. In a debate on energy in the House of Lords he permitted himself the statement that 'some of our most convenient fuels are going to be in short supply at an inconveniently early date' and when pressed went on to say: 'The trend curve for total energy consumption in this country is still following roughly an exponential law, but the different types of fuel used within that total have departed widely from the exponential laws that they were following. It is to accommodate such changes that one needs an energy forecast and an energy policy.'

Forecasting had to be a live, continuing activity, he said, and an energy plan would have to be flexible, unbiased, constantly under revision, and above all courageous. He supported such a plan, but politics would be so interwoven with it that it needed to be a minister's responsibility, although outside experts would also be necessary. To an audience of engineers,* he was even more direct:

> The best way of introducing greater accuracy into the forecast, would be to build power stations in 3½ years as they do in some overseas countries instead of 6 years as we do here; forecasting four years ahead is much easier than forecasting seven years ahead.

Dr Hubbert's forecast of exhaustion of fuels would not, in his view, bring economic pressures to bear before the end of the century – it ought to but human affairs were not like that – 'Posterity makes no contribution to dividends nor has it any votes in the next election. What other considerations enter into the determination of policy?'

* 'World sources of energy in the late twentieth century.' Graham Clark Lecture, 1971, Council of Engineering Institutions.

Lord Hinton is so quotable. Since I have already torn out from his logical and reasoned dissertation those pieces that I fancied to illustrate the theme here, in justice to him, but also for the greater benefit of my own crystal ball, I give his conclusion:

> Until recently the oil industry has been able to hold its prices at a remarkably stable level and it has been helped in this by the development of the giant tankers and by the discovery of new oilfields. Who is there who, 15 years ago, would have forecast the developments that have taken place in Nigeria and the Sahara, the recent developments in Alaska, the large proved gas reserves in West Pakistan or off the Australian coast or in the North Sea? How does one forecast developments up to the end of the century? Within the last year the oil-producing countries have shown increased firmness and solidarity in forcing up the price of crude oil; I believe that they will continue to do so. But, in spite of this, my global guess is that heavy fuel oil prices for the large industrial user will be comparatively (and I emphasize comparatively) soft. Crude oil prices have risen and will continue to rise over the years but my guess is that the oil companies may allocate a disproportionate part of these increases to the lighter fractions that are used by the tied transport and domestic users and that the price of the heavier fractions that are used in industry will be cushioned. Whether this happens or not will depend, mainly, on how effectively nuclear power competes with the fossil fuels and, in my opinion, it is outstandingly important that this competition should be effective.
>
> I have not burdened you with a lot of statistics because it seems to me that, by tying themselves too closely to statistics and by relying too little on objective arguments, forecasters have in the past distorted their crystal balls. What one can say with reasonable certainty is that up to the end of the century world energy requirements will continue to grow at a rate of around 5 per cent a year and that, in this period, the importance of mined coal will diminish and the consumer will increasingly demand 'piped energy', which means, in order of increasing convenience and sophistication, oil, gas and electricity. What seems to me to be uncertain is how well nuclear power will compete with oil in fixed land-based power plants up to the end of the century; in the interests of posterity I hope that it will compete strongly, but as I see things at present, I am not confident that it will.

To round off. The Western world plus Russia dominate completely in setting technical trends and in innovation, although Japan has shown good form recently in related technologies. Favourable new methods can

TABLE 7: WORLD ENERGY AND WORLD OIL

World energy usage, and usage in certain industrialized countries, expressed in million tonnes coal equivalent

World energy production	1950	1960	1970		
Coal and lignite	1,600	2,190	2,500		
Oil	700	1,400	3,050		
Natural gas	260	620	1,300		
Nuclear/hydro	40	90	200		
Total	2,600	4,300	7,050		
	1950	1960	1970	1980 (estimate)	1985 (estimate)
U.S.A.	1,230	1,620	2,450	3,700	4,500
E.E.C. (the Six)	300	470	850	1,410	1,810
U.K.	230	270	330	420	470
Japan	50	120	410	780	1,080
Total	1,810	2,480	4,040	6,310	7,860

World oil reserves, consumption and imports to certain countries, expressed in million tonnes of oil

	Proved reserves 1971		Oil consumption 1971	
	Million tonnes	% of world total	Million tonnes	% of world total
Middle East and Africa	58,000	66	99	4
North America, Western Europe and Japan	9,000	11	1,664	69
Rest of world	20,000	23	633	27
Total	87,000	100	2,396	100

Imports	1970	1985 (estimate)	Change
U.S.A.	157	740	+580
W. Europe	613	1,040	+430
Japan	199	530	+330
Total	969	2,310	+1,340

Conversion factors:		
	1 tonne coal equivalent	= 8 MWh = 30 GJ approx.
	1·5 – 1·6 tonne coal	= 1 ton oil
	75–80 M tonne coal equivalent	= 1 M barrels per day of oil
	1 barrel	= 35 imperial gallons
	1 tonne	= about 7·15 barrels

Sources: these are widely quoted figures, used here as collected by the British Coal Industry in *Coal and Energy Policy in Europe*.

generally be applied worldwide without great difficulty because of the need of all the western countries to export, and also because of the growth of aid to the less-developed countries, especially multilateral aid. Electricity supply is one of the economic spheres where multilateral aid can function effectively and on a large scale, accompanied by thorough appraisal, and this has been a factor accelerating the establishment and growth of competent systems of electricity supply in many countries.

Throughout the world then, a doubling in the size of electricity supply can be expected in about eight years. Other forms of energy usage will grow, with considerable variation but overall perhaps rather more slowly than electricity. Transport will be a dominant influence, and wherever new highways are opened, rapid increases in motor vehicles will be inevitable. This pattern is not so very different from the wider economic one which the world will see during that period, and indeed energy usage depends on the wider pattern.

One would hope that the somewhat frenetic attitudes and postures of some of the naturalists would settle down over the period into an attitude favouring responsible stewardship of the world's house-cleaning chores. There is no reason why standards of care for the atmosphere, the oceans and amenities generally should not improve greatly – what else is economic growth for? All this can be done without crippling the initiatives and the growth of the energy industries.

Less than adequate attention surely has been paid to the low percentage of utilization of all types of energy. The cost of the useful fraction is the practical criterion, but if prices rise, may I repeat what has already been stated once? Losses globally have been estimated at 80 per cent of the total. To reduce them to 60 per cent would accommodate one doubling in world demand. There appear to be no very dramatic ways of doing this; it would seem to depend on better practice and habits everywhere, as much as on innovations and techniques which enhance thermal efficiency.

The longer view

It is irrational to expect that life today should be ordered along lines which are stable and enduring on a geological scale of time. If the descendants of those alive today can be handed on a tradition of education and competence, in the widest sense, of managing and understanding the methods and devices of the world of energy, then as their lives run on, they will find answers to the inevitable problems that will arise. There are potentially other, more serious, perturbations to stability than the possibility of exhausting this or that commodity, even if it is a fuel. Nevertheless, there are reasons for looking as logically as possible at what is likely to lie ahead.

The first and most dominant thing is how many people will there be. In my view the best estimate to take is the World Bank's. If all methods of population control are assiduously applied as quickly as possible, projection A in Table 8 is the result. If a more likely, slower, rate – judged practicable in relation to what has already been done – happens, then projection B will result. Assuming the latter, and assuming also that the U.S. consumption in 1985 will be a reasonable levelling-off figure,* one arrives at a stable energy consumption of about 400×10^9 tons of coal equivalent, or nearly 60 times the figure for 1970, for a uniformly affluent world population over four times more numerous than today. This could happen by 2125. It is a very long assumption that it will. Energy usage depends on capital-intensive devices, to

TABLE 8: WORLD BANK'S POPULATION PROJECTIONS

	Population in billions				Stationary population in billions		
	World in years		Less-developed countries in years		Reached by about the		Less-developed
	1970	2000	1970	2000	year	World	countries
Projection A	3·6	5·9	2·5	4·5	2075	8·4	6·7
Projection B	3·6	6·7	2·5	5·3	2125	15·4	13·4

Projections assume:
A. Maximum effort in family planning plus other reasonable measures to curb fertility. A 'net reproduction rate' of 1·0 (a population with a net reproduction rate of 1·0 would continue to grow for 60 or 70 years until its age structure became stationary) would be possible by the year 2000 if the most successful results so far achieved became universal, but such a result is considered probably not realistic to expect.
B. Modest family planning as at present. A net reproduction rate of 1·0 would be reached by about 2040.
The table shows the populations that would be expected for each of these projections.

win fuel, to generate and transport energy, as much as to use it in motor cars, buildings and so on. There is an even larger assumption in increasing all those things by fifty or sixty times than is presented by the enormous rundown of fuel resources which such a consumption entails.

There is no impressively valid reason for assuming the levelling-off figure. It will depend on the philosophies and motives which guide the lives and mould the purposes of the numerous population of the twenty-first and twenty-second centuries. They will be healthy and well-fed, one assumes – which is not the case today – but will they be uniform in their habits, like the worldwide box-like uniformity of twentieth-century architecture? Or will the Masai still roam undisturbed in East Africa, and the suburban Londoner still commute in daily routine? Will there be a competitive urge to be better off than one's neighbour, or nationally better off than the next country? A levelling-off in energy consumption presupposes a levelling-off in economic growth, which in a society rather substantially better off than the U.S.A. today would seem to

* If the *per capita* consumption of energy of the whole of the world's population were equal to the *per capita* figure in the U.S.A. world consumption today would be about 6·8 times larger than it actually is. World energy consumption in 1970 was about 7×10^9 tons of coal equivalent. At the U.S. figure for 1985, this would increase to $7 \times 6{\cdot}8 \times 2 \times \frac{15{\cdot}4}{3{\cdot}6} =$ approx 400×10^9 tons of coal equivalent.

depend on that basic urge becoming perhaps unfashionable, or even immoral.

There is certainly much still to be done, notably in the less-developed countries, and much incentive to 'catch up', but it is a moot point how much more wealthy and extravagant the richest country of the world, whether it remains the U.S.A. or not, will become. Maybe in another century, having pulled up the poorest countries to acceptable standards (as measured in liberal terms by the western world today) the world might well be moving to a different social philosophy and discarding the industrially-based acquisitive habits of today. A world peopled by healthy, well-educated people indifferent to personal wealth would be fine. But the decadence of comfort is difficult to discard. Education today demands travel and expensive gadgetry; the world has learned to suck life like an orange, rather than to serve, for the glory of it, without thought of reward. People don't change. So if some latter-day sylvan idyll came to pass, it would be surely be accompanied by energy consumption at least as high as today's peaks. These homespun speculations are aired because energy is the servant of wider purposes.

A sequence of possibilities

The possibilities of exhausting fuels can be examined in sequence, assuming first, success of some particular method, and then that it will fail and an alternative will have to be found. This makes explanation easier to follow, but is not likely to resemble what will happen in the world at large, where all methods, with varying success in different places, each tend to have a share.

First then, if the fusion reaction is controlled and utilized, and developed to exclude lithium, there will never be an energy famine, strikes, war, politics and mismanagement excepted. Chapter 5 explains the prospects. It sparingly quotes a statistic or two for the virtually infinite resources which this process would make available, and points out that the process itself is likely to be cumbersome because of its great scale, but fundamentally benign and more nearly harmless than we have any right to expect from so potent a source of energy. Even with barely any other fuels at all, transport could depend on batteries recharged from the electricity grid or on hydrogen made by electrolysis, and society could continue in its customary wasteful way.

Suppose then, failure to control fusion. Breeder reactors, at a cost of producing radioactive wastes, some of long life, and the risks inherent in having an explosive quantity of radioactive, toxic fuel in every power station, could function on known and proved supplies for more centuries than I need to specify. Dr Hubbert says that U.S.A. sources of uranium alone, used in this way, would be a thousand times greater than all the world's fossil fuels.

This life is perhaps fifty times greater than can be proved now, for the slow neutron reactors that are now in use. So if the fast reactor failed, or became unacceptable, present nuclear fuels, already proved, and used in fission as at present practised for power generation, would suffice for a while, but a finite while, longer than but comparable with the life predicted for fossil fuels.

Oil is the fuel around which discussion tends to centre. When will oil run out? Its consumption has grown so fast, despite the fears which have always been voiced, that it is near the point of exhaustion. Proved reserves in 1950 were sufficient for 80 years; proved reserves now are sufficient for 30 years. Dr Hubbert suggests that with a doubling of oil consumption every 8·4 years (1930 = $12{\cdot}3 \times 10^6$ barrels) oil will reach its maximum production between 1990 and 2000, and then decline, being virtually exhausted by the late twenty-first century, but this estimate excludes oil from tar- or heavy-oil sands.

There are reasons for uncertainty about future oil supplies. It is expensive to prospect and to prove resources. It pays to keep quiet about future sources in some circumstances. Politics, rivalry, opportunism and profit combine to distort an objective appreciation of the world's resources of oil. There is yet another factor: ignorance.

One of the most fascinating reports of scientific investigation that has been published recently is that of the voyage of the Glomar Challenger. This ship was equipped to drill the Mohole, a hole so deep that it would penetrate the Mohorovicic discontinuity (i.e. the base of the earth's crust) and discover what the inside of the earth was made of. It never did that job; difficulties and costs conspired to defeat the main aim of the project. Instead, the Glomar Challenger went on a cruise and by mid-1973 had drilled over 400 deep holes at chosen points around the oceans of the world. She was equipped to drill in depths of water of three miles, and the record depth of water quoted is just over 20,000 feet. The maximum depth of the hole drilled in the sea bed at that depth is not stated, but one exampled quoted is over 4,000 feet, beneath 11½ thousand feet of water. The published report contains a wealth of fascinating knowledge, mainly concerning the geological structure of the sea bed and the theories of geological evolution deduced from these observations. The report is less explicit – though still with sensational hints – about commercially valuable deposits of metallic ores and minerals, of which it found plenty. It is, one would guess, censored about the possibilities of new sources of oil, bearing in mind that one of the sponsors was a U.S. oil company. But the report drops intriguing suggestions. Oil was found at depths greater than thought possible. There seems to be plenty of oil and gas beneath the Gulf of Mexico, and beneath the Mediterranean Sea there could be the biggest oilfield of all time.

In the shade of the gigawattleia *tree*

Consider, if nuclear technology either failed or were rejected, and if oil and coal ran out, or substantially so, and oil shales and tar sands too had been exhausted. What then? Hydro-electric power would be extended to its limits throughout the rivers of the world, maintaining its present proportion of about 2½ per cent of world consumption of energy until, say, at least forty times present demand. Some tidal power would be generated, and tidal currents harnessed. Wind, with all its fickleness, would be extensively harnessed. If tides and winds contributed equally with hydro-electric power up to the multiplying factors of ten or twenty, and residual fuels doubled the total amount, they would all still have met only a fifth or sixth of demand. The direct exploitation of solar energy would expand dramatically in tropical zones, but suppose that too had rather severe limits. Geothermal power similarly. In a world that rejected nuclear power, the cataclysmic explosions of Rougeron, even if they worked, would be taboo, and so the gross output would be limited by being confined to the exploitation of natural sites.

The technologies of space might perhaps be rather advanced, to a point where those planets which have atmospheres of light gases, in particular methane, could be exploited. Gas could be transported to earth, either in capsules (astro-L.N.G.) or by some transplanetary gaseous jet devised by a twenty-first-century physicist, so giving an inexhaustible source of fuel.

Maybe that would be too complicated. There is a simpler alternative. If the twenty-first-century physicist let the world down, then the twenty-first-century horticulturalist would come to the rescue. Wood is a traditional fuel. He would discard it as too primitive, and devise and cultivate a plant – tree or vegetable or a sort of Jack's beanstalk – which grew rapidly and yielded, as timber or fibre, copious quantities of fuel. The aim would be to grow not two blades of grass where one grew before, but ten joules where one grew before.

In a good year a crop of about 80 tonnes of tomatoes can be expected today from an acre (4,000 m^2) of heated glasshouses; the figure tends to improve as the years pass and methods improve. If one assumes that the new fuel plant – say it would have pretty blossoms and we can call it *gigawattleia* – yields a crop which weight for weight has half the calorific value of coal but can be grown at twice the tonnage of a good tomato crop today, then the whole of the energy demand of Britain for about the year 1980 could be met from cropping 200 by 100 kilometres of land. Irrigating the Sahara would supply the world with energy, and restore OPEC to its lost peak of influence. A world energy demand of 8×10^9 tonnes of coal equivalent (about the present one) could be grown on a square of land about 650 by 650 kilometres. Oil would be turned into protein and eaten, and *gigawattleia* grown all over the place. However, tomatoes are somewhat watery and this yield in terms of solar energy is about a thousand times greater than Professor Wolf suggests 'as a rule of thumb', but it is only about 7 per cent of the total solar energy that might be expected in a year at a rate of 0·8 kW/m^2 for 8 hours a day.

The increased quantity of green crop maybe would compensate by photosynthesis for increasing concentrations of CO_2 in the atmosphere, but if not a synthetic photosynthesis (sic) plant could be devised, presumably using the sun's rays directly.

To a rude mechanical, growing *gigawattleia* seems more practicable and more easily assimilable in the mainstream of human affairs than fussing with lagoons full of slime, which is the alternative suggested for utilizing solar energy, but life is full of surprises.

Doom and damnation versus cheerful home fires

Global computer models, such as the Club of Rome's, and the dire forecasts from them, have been debunked rather thoroughly in learned circles (for example with 12 variables each having three possible values 3^{12} answers result) but maybe it is worth mentioning some of their limitations, apart from the inaccuracy of the data from which the predictions start. The argument is virtually entirely statistical. It ignores the biggest regulating factor of all, which is price – when a fuel begins to get scarce its price rises and adjustment begins. It also ignores inventiveness – a fault in common with all predictions including the ones in this chapter – and prejudges exploration. Further, it ignores the flexibility of modern industry and technology, which has considerable powers of adaptation should difficulties or the unforeseen arise.

Other pessimistic forecasts stress the universality of the laws of thermodynamics and suggest that, since all energy usage degrades into the dissipation of heat, eventually the earth's average surface temperature will rise. One source suggests that this will happen when man's energy grows to 1 per cent of solar input (presumably gross input – 5 per cent not 1 per cent was quoted in Chapter 6 from another source, which gives an idea of the 'factor of ignorance') which is somewhere around three times the levelling-off figure quoted in this chapter; then the polar ice caps would start to melt. This is a rather hasty conclusion. By that time the terrestrial heat engine – the why and how of weather – may well be understood. Such knowledge would add immensely to man's potential. The average temperature is not so critical as the variations (as the man said 'it's when it ain't average that bothers me!') and is it now the right average? One can envisage perhaps half-a-dozen man-made gulf streams utilizing the earth's energy of rotation (but not extracting from it, naturalists note, to slow the earth down) to promote mild and healthy climates where now it is bleak and uninhabitable; even ice and solar gain can be controlled. The optimistic speculation is neither more nor less idle than the pessimistic one.

My crystal ball, then, says, Don't worry. Some genius contemporary with your great-grandchildren will sort out any difficulties if any should arise by then. But if the human race does not blow itself up in a fit of anger or jealousy, why should not the purveyors of energy continue to ply their business diligently? In the meantime energy could even get as expensive as beef, but in this affluent world you will still be able to afford more of it, and waste 80 per cent of it too.

Nuclear power was unknown as a practical influence in human affairs until 1945; its theoretical basis was evolved remarkably quickly in the years before that date. Fusion is even younger, and has perhaps just reached the stage where it is considered a reasonable prospect. For the future, I quote Sir Brian Flowers*:

> Could it be that a massive reappraisal of physical laws, a new synthesis perhaps, could come to pass in time to affect substantially the practice of engineering in the year A.D. 2000? It is just possible. Are there any signs of it happening? There are; for nuclear physicists have perceived a new pattern of highly energetic phenomena in the subatomic world, and astronomers are already speculating on whether this pattern has any bearing upon the huge energy sources (the quasars) recently discovered in the distant universe and on whether the gravitational forces of such dense objects interact with, and modify, the other forces of nature. There is a hint of a new synthesis in the air. It has to do with esoteric and unworldly phenomena; but so did the quantum theory of the 1920s which now pervades all of science and much of technology.

* Professor Sir Brian Flowers, chairman, U.K. Science Research Council, 'The scientific basis of electrical engineering in AD 2000', Institution of Electrical Engineers centenary lectures, 1971.

Appendix

The unique nature of heat

The first law of thermodynamics has to be qualified because of the unique nature of heat. There is always equivalence between the different kinds of energy, but there are some limits on convertibility. A given quantity of heat cannot be converted entirely into mechanical or electrical energy. The idea of potential energy has already been introduced. 'Potential' in the case of thermal energy is temperature. High temperature gives a high grade of energy; some low-grade energy cannot be converted, and the efficiency of conversion depends on the quality of the thermal energy. Mechanical and electrical energy are of uniformly high quality, and can be fully converted (in theory) into the other forms. Heat is thus the lowest form of energy, and its quality depends on its temperature.

An analogy may help to explain why only part of a given quantity of heat can be converted into work or electrical energy. Suppose one's source of energy is a set of 1 lb weights, positioned 10 feet above a tank of water, 5 feet deep. The potential energy of each weight, measured to the bottom of the tank, is 15 ft lb. However, it is possible to utilize only the energy above the water line, i.e. 10 ft lb, and the remaining 5 ft lb is dissipated as the weight drops uselessly through the water. For a weight held 10 feet higher, the proportion of its potential energy lost is smaller.

The second law of thermodynamics

Conversion of thermal energy into work involves degrading it to a lower temperature (equivalent to the water level in the analogy) below which heat is rejected. The proportion which can be converted into work depends on the range of temperature and is expressed by $\frac{T_1 - T_2}{T_1}$ where T_1 and T_2 are the high and low temperatures respectively, expressed in the scale of absolute temperatures (that is in kelvins as defined a little farther on). Absolute zero is -273 on the Centigrade scale, so for a conversion in which the high temperature is the boiling point of water and the low temperature the freezing point of water, the maximum efficiency of conversion theoretically obtainable is $\frac{373 - 273}{373}$

or about 29½ per cent. If the maximum temperature is changed to 1,000 deg.C, the maximum efficiency is $\frac{1273 - 273}{1273}$ or about 78 per cent. The proportion of heat not converted is dissipated. It follows that there is a running down of high-grade energy in every conversion from heat to work. Another way of expressing the second law of thermodynamics therefore is the statement that it is impossible to convey heat from one body to another which is at the same or a higher temperature, without the aid of external energy (a refrigerator consumes power in order to 'pump' heat from a low temperature to a higher, ambient, temperature).

Carnot's reversible cycle

Sadi Carnot discovered the second law of thermodynamics in 1824. A law of this kind was sought to express in a scientifically formal manner the impossibility of perpetual motion, and the law has extraordinarily far-reaching implication in scientific thought because it implies that there is a tendency for systems (e.g. the universe) to run down.

Carnot, however, was concerned with determining the maximum efficiency which in theory could be obtained from a heat engine, i.e. a device which converts thermal energy into mechanical energy. A heat engine must have a source of heat at the high temperature at which it works (T_1 in the expression above), a working agent, and a sink at the low temperature (T_2 above) to which it rejects heat. Carnot said that irreversible changes which occurred in the cycle of operations of a heat engine caused losses of energy; so the cycle of operation of his perfect heat engine is fully reversible. The working agent in such an engine is an ideal or perfect gas, working in a cylinder and with a piston of non-conducting materials. In other words, losses of heat to the surroundings or of energy by turbulence of the gas, are assumed not to take place. The working cycle of the engine consists of, first, an isothermal (i.e. at one temperature, in this case T_1) expansion, during which heat is abstracted from the hot source, and the piston moves outwards in the cylinder; in the second stage the source of heat is cut off; expansion continues, but adiabatically ('adiabatically' means that the total energy of this stage remains unchanged). At the limit of the piston's stroke, the third stage, isothermal compression, commences; during this stage heat is rejected to the sink at the low temperature, T_2. In the fourth stage compression continues, but adiabatically, so that when the piston has completely returned, the working agent has regained exactly the same temperature and pressure as at the start of the cycle.

Fig. 21 below shows the Carnot cycle; it is a sketch familiar to all students of thermodynamics. Pressure is plotted against volume, in the

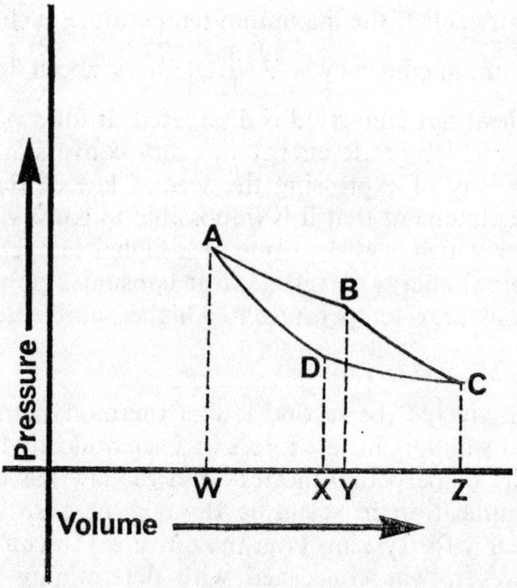

Fig. 21 Carnot's cycle.

form well known to engineers as an 'indicator diagram' commonly derived from tests of reciprocating engines. If pressure is plotted against volume in this manner for the working cycle, the work done in the cylinder is measured by the area traced out in the diagram. In the case of Carnot's ideal cycle, the heat supplied during the first stage of isothermal expansion is equal to the work done during that stage and is ABYW. In the second stage, the total energy remains constant; as the gas expands adiabatically, it does work, measured by the area BCZY, and its internal energy decreases by that amount – it gets colder. In the third stage, as it is compressed isothermally, work is done on it, and it rejects heat, measured by CDXZ, to the sink. In the fourth stage, which is adiabatic compression, work equivalent to the area DAWX is done on the gas and its internal energy increases by that amount. The net result of the cycle in terms of useful work done is thus the area ABCD, the heat supplied is ABYW, and the heat rejected is CDXZ. Analysis of these stages gives $(T_1 - T_2)/T_1$ as the efficiency of the cycle (efficiency = work done divided by heat supplied).

Carnot's engine is reversible, and if the cycle is considered in reverse, work equivalent to the area ABCD is done on the gas, and heat abstracted from the sink at temperature T_2 and delivered to the hot source at temperature T_1; the cycle in reverse is thus the most efficient performance that can be expected from a refrigerating machine. Formal proof that this reversible cycle is the cycle of greatest theoretical efficiency is given by considering two such engines working between the same source and sink, one as an engine and the other in reverse as a heat pump. If the one working as an engine were more efficient than the Carnot cycle, heat would be transferred to the source at the higher temperature without the aid of energy external to the system, which would be contrary to the second law of thermodynamics and, like other postulates which are contrary to that law, would be an example of perpetual motion.

This background of thermodynamic theory has been sketched out to illustrate that heat or thermal energy, while it is strictly equivalent to the other forms of energy in numerical terms, is energy of lower quality. When it is converted to the other forms, a certain proportion of it is lost. This limitation will be with us constantly, as we explore the uses of engines and the generation of electricity.

Units of energy

Having emphasized that the various kinds of energy are quantitatively convertible it might be expected that they would all be measured by the same unit – for example the foot-pound, which is the only unit mentioned thus far. One could use the unit, say the foot-pound denoted by FP, with a suffix or some distinguishing mark if the type of energy had to be specified, for example FP_h for heat or FP_e for electrical energy, and the one unit would suffice for all purposes.

History shows that exactly the opposite has occurred in practice; each type of energy has its own unit or group of units for each common system of measurement. Although Britain is now using the SI system of metric measurement, British people still think in terms of the Imperial system, now discarded at least officially, and probably nearly everybody knows instinctively just what is a foot or a pound, but maybe not, without thinking it out or referring to a book, just what is a kilogramme or a millimetre. The metric system as used by continental countries is not necessarily the SI system, and may differ appreciably from it; the Imperial system, with some variation from its British prototype, is in use in the U.S.A. So according to the immediate needs of practice, and following habits so developed, a great variety of units of measurement has evolved; the variety is as great as anywhere in the subject of energy. Variety can be tolerated, but confusion cannot, so the purpose here is to give some guidance in the form of definitions of the commonest units and the relationships between them. Having done this, some idea of the physical significance of common units can be imparted to the reader. This again is a useful preliminary, since size and capacity are important characteristics, and with this knowledge it becomes easier to compare and to evaluate figures when they are quoted.

The fact that energy, in its various manifestations, is a basic phenomenon of the physical universe was first illustrated by the difficulty of defining it in any simple way; the same fact is again illustrated by the variety of units in which it has been necessary to measure it.

The SI system of units

The SI (*système internationale*) was not evolved until 1960. SI units have been used mostly in this book, but how they relate to familiar habits of measurement has to be made clear. First of all, here are the five primary units which are the basis of all physical measurement in the SI system, and the symbols which are used to represent them:

Unit of:	length	mass	time	electric current	temperature	luminous intensity
Unit	metre	kilogramme	second	ampere	kelvin	candela
Symbol	m	kg	s	A	K	cd

For the next step, we will return to an old friend, the foot-pound, before leaving him behind along with the Newcomen engine and other ancient relics.*

In the old systems, mechanical work is measured in foot-pounds or in kilogramme-centimetres, or in foot-poundals or dyne-centimetres (ergs) if the gravitational constant (g = 32·2 ft/sec^2 or 981 cm/sec^2) is avoided in the definition of the unit. Thus: 1 ft lb = 32·2 ft poundals = 13·83 cm kgs = 1·356 × 10^7 ergs.

The unit of force in the SI system is the newton, which is the force which will accelerate a mass of one kilogramme by one metre per second per second (i.e. by approximately one tenth of the gravitational constant of 9·81 m/s^2). One newton acting through a distance of one metre is the unit of energy and is called a joule; work, and energy in all its forms, are measured in joules – the logical answer we were led to expect.

Power is measured in joules per second; one joule per second is one watt. Here then is the system of nomenclature and symbols for the SI units of energy and power:

* Recent practice in scientific and technical publications has been to write 'lbf' for pounds, meaning 'pounds force', e.g. one foot-pound is written 1 ft lbf. The reason is that a force of one pound is exerted when the earth's gravitational field acts on a mass of one pound; mass is constant, but gravity is not, a fact made explicit by the exacting pedantry of this notation. For most of us, our scientific integrity is not put in jeopardy by conveniently omitting the distinction for practical purposes, but apparently the permissive society permits even pernickety usage of scientific symbols.

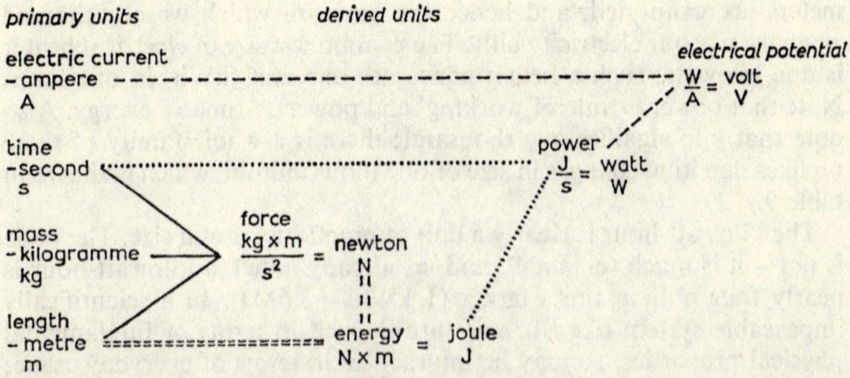

Rearrangement and adaptation of units already familiar is apparent in this scheme. Electrical units fit into it unchanged for all practical considerations. The watt, the unit of power, is used frequently throughout this book, but as can be seen, the intention of the SI system is to use it universally; for mechanical power, the horse-power becomes obsolete, and is replaced by the unit already familiar in electricity. Similarly, the rate of flow of heat and refrigeration (British thermal units per hour and tons respectively in the Imperial system) are expressed in watts.

Units of heat have disappeared – the British thermal unit (the heat needed to increase the temperature of one pound of water by one degree Fahrenheit) and the calorie (the heat needed to increase the temperature of one gramme of water by one degree Centigrade). But the SI system makes some concessions, albeit small ones, to common usage. It 'permits' one unit we use frequently, the kilowatt-hour (one thousand watts for one hour = 3·6 million joules), the use of minutes, hours and days to measure time, and the tonne (a thousand kilogrammes, and almost equal to a ton) to measure mass; but there does not seem to be a unit of *noblesse oblige* in the SI system, and practice defers to science.

The physical size of common units

That is the system. Let us grasp its meaning in comfortably familiar habits of perception. Energy being our subject, we will concentrate on the joule. A joule is the energy expended by a force of one newton (about one tenth of a kilogramme or a quarter of a pound) acting through a distance of one metre (one yard plus $3\frac{1}{2}$ inches). Which works out at about three quarters of a foot-pound: more exactly,

$$1 \text{ ft lb} = 1{\cdot}356 \text{ J} \quad \text{and} \quad 1 \text{ J} = 0{\cdot}7376 \text{ ft lb}.$$

The kilowatt-hour is very familiar as the 'unit' in which electricity meters are calibrated, and hence the basis on which we are charged when we pay our electricity bills. The commonest size of electrical heater is one kilowatt; such a heater consumes one unit (kWh) in one hour. Note that power is 'rate of working' and power × time = energy. Also note that kilo signifies one thousand; there is a whole family of these prefixes denoting changes in size of one thousandfold, which is shown in table 9.

The kilowatt-hour is clearly a unit of practically useful size. The joule is not – it is much too small, and, as already noted, a kilowatt-hour is nearly four million times larger (1 kWh = 3·6MJ). In a scientifically impeccable system like SI, units are defined in terms of fundamental physical properties, so may be impractical in terms of everyday usage; hence the battery of × 1000 prefixes. However, there is no great difficulty in replacing horse-power by kilowatts; the conversion is:

1 hp = 746 W (0·746 kW)
1 kW = 1·341 hp.

Heat tends to be needed in larger quantities than other forms of energy, perhaps as a consequence of its lower quality. A one-kilowatt heater is quite modest in the duty it does, but a light bulb of one-tenth of that power (100 W) gives a very bright light. A washing machine may be operated mechanically by a motor of perhaps 300 W or 500 W, but it will need a heater of three or four times greater power. Table 9 also lists common units that are applicable to our subject, and gives conversion factors. In Chapter 2 some examples are given which indicate the physical significance of some of these units in terms of outputs and consumptions, as we find them today. This is the end of the 'pill' section, and the reader should now be equipped to find out why and how he is responsible for enormously running down the resources and quality of energy on this planet.

TABLE 9: UNITS OF ENERGY: CONVERSION FACTORS: THOUSAND-FOLD PREFIXES

Unit	Definition	Conversion factor
Joule, J	One newton acting through one metre – the SI unit of energy	—
Kilowatt-hour, kWh	One thousand watts for one hour	1 kWh = 3·6 MJ
Horse-power-hour; hph	One horse-power (33,000 ft lb per minute) for one hour	1 hph = 1,980,000 ft lb = 2·685 MJ
British thermal unit, Btu	Heat needed to raise the temperature of 1 lb of water by 1 deg Fahrenheit	1 Btu = 778 ft lb = 1055 J
Therm	100,000 Btu	1 therm = 29·3 kWh = 105·5 MJ
Note: 1 tonne coal = 230/320 therms; 1 gal oil = nearly 2 therms		
Calorie, cal	Heat needed to raise the temperature of one gramme of water 1 deg Centigrade. The calorie used in dietetics is one thousand times larger	1 cal = $4{\cdot}19 \times 10^7$ ergs = 4·19 J
1 tonne coal-equivalent	Energy of any kind equal to the heat generated by burning a tonne of coal of average calorific value (a high value could be 15% more)	1 tonne coal equivalent = 8,200 kWh = 30 GJ approx. = 0·6 tonnes oil = 280 therms
Foot-pound, ft lb	As defined in the text	1 ft lb = 1·356 J
Foot-poundal	Ditto	1 ft-poundal = 0·042 J
Centimetre-kilogramme	Ditto	1 cm-kg = 0·098 J
Erg (dyne-centimetre)	Ditto	10^7 ergs = 1 J
Electron volt, eV	Energy given to an electron by one volt	1 eV = $1{\cdot}6 \times 10^{-12}$ ergs = $1{\cdot}6 \times 10^{-19}$ J

THOUSAND-FOLD PREFIXES

1 000 000 000 000	$= 10^{12}$	= T (tera)
1 000 000 000	$= 10^{9}$	= G (giga)*
1 000 000	$= 10^{6}$	= M (mega)
1 000	$= 10^{3}$	= k (kilo)
0·001	$= 10^{-3}$	= m (milli)
0·000 001	$= 10^{-6}$	= μ (micro)
0·000 000 001	$= 10^{-9}$	= n (nano)
0·000 000 000 001	$= 10^{-12}$	= p (pico)

* Pronounced as in gigantic. Note the important distinction between M and m.

TABLE 10: FUELS, SOURCES OF ENERGY AND METHODS OF CONVERSION

Fuel or source	*Method of conversion or utilization*
The sun's radiation	Natural processes (weather; photosynthesis) Direct use (mirrors; solar cells; evaporative hydro-electric)
Weather: wind	Turbines and electricity generation
water (rivers, sea currents, tides)	Turbines and electricity generation; heat exchange from warm waters
Food	Muscles
Geothermal heat	Heat engines and electricity generation; direct use
Fuels:	
dung; animal fats and oils; wood; peat; other crops; lignite; coal; oil and natural gas; oil shales; tar sands; natural bitumens; chemical fuels, natural or derived, e.g. alcohol, hydrogen	Heat engines; then direct use of mechanical energy, or conversion mainly electricity generation, including complex, multiple-stage or multiple-purpose processes The fuel cell; batteries; explosives; the heat pump.* Direct use of heat.
Energy of the atomic nucleus:	
fission; fusion	Explosions; controlled transmutation of mass into energy, generally heat energy, then utilization as for fuels

Direct conversion of energy into electricity

Fuel cell: chemico-electric generation as in a battery. The reagents are continually supplied from outside, and the fuel cell does not exhaust itself like a battery.

Thermo-electric generators: thermo-elements from semi-conductor materials.

Magnetohydrodynamic generators: electric currents are induced according to Faraday's laws in an expanding electrically-conducting gas by strong magnetic fields, and are conducted away as useful current.

Thermionic generator: a current is generated at the colder anode by thermionic emission from a hot cathode.

These four definitions were given by Dr Jenny in his general report to the fourteenth sectional meeting of the World Energy Conference.

* Refrigerator or heat pump: A working agent is compressed and condensed using mechanical energy from an external source. At this point in the cycle the working agent rejects heat at the high temperature. The working agent then expands and evaporates and absorbs heat at the lower temperature. As a heat pump, the heat rejected at the high temperature is utilised. As a refrigerator, the heat abstracted at the lower temperature is utilised.

TABLE 11: A COMPARISON OF ENERGY NEEDS AND OUTPUTS

Note: 1 All conversions into kWh are gross figures
2 't coal equivalent' means the energy released when a tonne of coal is burned (see table 9 for conversion)
3 't' means the energy released by the explosion of one tonne of chemical high explosive (TNT) and is converted below

Category	*usual unit*	*converted to kWh*
One man: one day's physical work	—	0·5 to 1·0
one year's physical work	—	200
one lifetime's physical work	—	10,000
one day's intake of energy (av. for Britain, 1972): food	3090 kcals	3·8
alcoholic drink	139 kcals	
one year's intake	1·2 G cals	1,400
one lifetime's intake	80 G cals	92,000
gross lifetime efficiency for physical work	about 10%	
The author's domestic electricity consumption for three winter months: main tariff	—	1,343
off-peak (heating)	—	5,959
solid fuel burned in same period	1 tonne	8,200
total domestic energy for three winter months		30,000
		coal equiv.
One car journey, about 40 kilometres on flat ground	1 gallon petrol	41
Longannet power station, Scotland (Chapter 11) 2400 MW installed:		
one hour's output at full load	—	2,400,000
one year's output at 67% load factor	—	14,000,000,000
corresponding consumption of coal	5×10^6 tonnes	—
Refining one tonne of steel	—	13,000
Refining one tonne of aluminium	—	70,000
Exploding one tonne of chemical high explosive (TNT)	1 t 4×10^{16} ergs	1,100
Explosion of an atomic bomb	20 kt	22,000,000
Explosion of a hydrogen bomb	50 Mt	55,000,000,000
Condensing one kilo of steam to water at 100 deg.C	535 kcals	0·65
Dropping one kilo through 100 metres	980 J	0·00027
Energy of fission: initial burn-up of one kilo of natural uranium in a slow reactor	10 t coal equiv.	80,000
processing fission products and recycling, from same kilo of uranium, total amount	30 t coal equiv.	240,000
theoretically complete burn-up of one kilo of uranium	3,000 t coal equiv.	24,000,000
Energy of fusion: fusion of all atoms of one kilo of deuterium	2,800 t coal equiv.	22,000,000
fusion of one kilo of reactants with lithium-uranium blanket	9,000 t coal equiv.	75,000,000
Complete transmutation of one kilo of matter into energy	3×10^6t coal equiv.	25,000,000,000

TABLE 11 – *continued*

Category	*usual unit*	*converted to kWh*
Energy in natural phenomena		
Energy radiated by the sun in one year	1·2 × 10^{41} ergs 5 × 10^{23} h.p.	3·3 × 10^{27}
Gross energy received by the earth from the sun		
in one year	4·4 × 10^{31} ergs	1·2 × 10^{18}
rate of direct intake	0·8 to 1·0 kW/m^2	
which for one year at 8 hours a day gives	0·3 t coal equiv./m^2	2300/m^2
Heat lost each year from the earth's interior	8 × 10^{27} ergs	2·2 × 10^{14}
Kinetic energy of rotation of the earth	2·6 × 10^{36} ergs	7·2 × 10^{22}
Annual release of energy by earthquakes	9 × 10^{24} ergs	250,000,000,000
Range of energy of individual earthquakes:		
largest (magnitude 9)	$10^{24·75}$ ergs	160,000,000,000
threshold of damage	$10^{18·9}$ ergs	220,000
Energy released by condensation of water in a typical thunderstorm (US)	200 kt	220,000,000
Energy of rotation of a North Atlantic Depression	250 Mt	300,000,000,000
Energy released per hour in a tropical hurricane	400 Mt	440,000,000,000
Energy needed to melt the Antarctic ice cap	2½ years' energy from the sun	3 × 10^{18}
Energy needed to raise the temperature of the oceans by 1 deg.C	ditto	3 × 10^{18}
Estimate of the energy of a supernova explosion	10^{51} ergs	2·8 × 10^{37}
World energy production (gross): 1968	6 × 10^9 t coal equiv.	49 × 10^{12}
1970: coal and lignite	2,500 × 10^6t	
oil	3,050	
natural gas	1,300	
nuclear/hydro	200	
total	7,050 × 10^6t coal equiv.	58 × 10^{12}
1985	12/16 × 10^9t coal equiv.	100/130 × 10^{12}
2000	30 × 10^9t coal equiv.	250 × 10^{12}
Overall worldwide losses in utilization	80%	80%
Energy consumption per year in Britain (1971)	323 × 10^6 tonnes coal equiv.	2·6 × 10^{12}
Energy consumption in the U.S.A. in 1970:		
total (all sources)	70 × 10^{15} Btu	21 × 10^{12}
per head	3·3 × 10^8 Btu ten lifetimes physical work	100,000
rate of growth forecast till 1985	4·2%	4·2%
consumption by purpose: industrial	40%	40%
transport	30%	30%
domestic etc	30%	30%

Note: the US consumes 35% of world energy, and has nearly 7% of world population

World consumption of fuels in the past century:		
coal	10^{11} tonnes	8 × 10^{14}
oil	10^{11} barrels	2 × 10^{14}
natural gas	10^{12} cubic metres	1,100,000,000

Postscript: For richer, for poorer

I finished writing this book in the summer of 1973. This short chapter is a sort of postscript, written in March, 1974. I have left the text unaltered, with just a note against those passages that have been superseded by events. So the reader will easily be able to see how events distort forecast or prophecy, in a subject that changes rapidly and unexpectedly in some of its aspects, but has a timeless quality in others.

Nuclear power

The fast breeder reactor is now crossing the threshold. If there are no setbacks in the immediate future, it will establish itself as a successful method of generating electricity. Fast breeder reactors in Britain, France and Russia have gone through the initial stages of trial and operation successfully and as expected. However, none of the advanced gas-cooled reactor (A.G.R.) power stations in the British programme are yet in service. There is ubiquitous pessimism about the future of the A.G.R.; one gets the impression that it will be fortunate if these stations can be made to work at a fraction of their rated capacity.

A secretary of state for energy

Early in 1974, proponents of an energy policy in Britain had a major success – a cabinet appointment, Secretary of State for Energy, was created. So far since the election, the post has been retained, but it is too early yet to see whether the quality of guidance that the proponents of that idea sought will in fact be given.

North Sea oil

The oil beneath the North Sea is seen as a golden cure-all for Britain's economic illnesses. Estimates of the yield from the North Sea continue to increase – the whole of Britain's needs in 1980 is an official estimate now – and the first oil is being hastened along as fast as tongues can hasten it.

A gush of money

The major events that made this postscript necessary were the Arab–Israel war and the subsequent use of oil as a political weapon, combined with drastic increases in its price. The power of OPEC was demonstrated rather sooner and more forcefully than had been expected. The oil companies are no longer as unique and select as they were described in this book, but have become virtually distributors. The sequence on page 86 has been overtaken by events and the OPEC states have taken firm control of their oil. They increased their revenues by a factor of four in two massive price increases in October and December 1973. The four-fold increase becomes something of the order of a 50 per cent increase in the price of the product as sold to the consumer, with certain qualifications. So reverting to page 142, perhaps half of the increase in the price of oil forecast until 1985 has already occurred, as far as supplies from OPEC sources are concerned.

Bearing in mind the principles discussed in earlier chapters, it seems reasonable to conclude that an increase of the order of 50 per cent is unpleasant, but not, in relation to a national economy, unbearably serious. Broadly that is true, and it is not the increase in price of itself that has caused the major worry. It is the problems of balance of payments, and the potential instability threatened in the international monetary system that are the most serious aspects.

Alternative possibilities have already been stimulated by these high prices, and the first auctions of mining rights for the oil shales in the Rockies have taken place in the U.S.A. Stimulation can take differing forms however. On this side of the Atlantic, its expression was a massive pay claim by British miners when oil was short, which led to the three-day working week, the miners' strike, and the general election. How much of that belongs in a book on energy? All these events, sadly enough, confirm one notion already made familiar in earlier chapters, that it is the motives and conduct of people that dominate human affairs, including the supply of energy.

Index